An Introduction to Einstein's Relativity
Version 1: Math Included

W. Blaine Dowler

April 3, 2025

Cover design by: Art Painter
Library of Congress Control Number: 2018675309
Printed in the United States of America

About This Book

The contents of this book were originally published in a series of nine weekly lessons at `http://www.bureau42.com` in the summer of 2012. Those lessons were released in two versions, as is this book. One version, the version you are holding, includes all of the math. The other, which you are less likely to be holding, omits the math. All of the content from the mathless version is present in this version, so you are not missing anything by picking up this version instead of the other one. Furthermore, this version cleans up some explanations and corrects some typos from the original web-based version.

Contents

Chapter 1

The Need For Relativity

1.1 The Success of 19th Century Physics

By the end of the 19th Century, physicists were standing proud. It seemed as though every unanswered question in physics was on the verge of being answered, and that physics would be the first field of science to be described in totality. The list of unanswered questions was short, and the answers seemed to be a matter of applying basic mathematics to several ideas and compiling the results. We shall now explore some of those specific questions in detail.

1.2 Mercury's Orbit

Most of us imagine planetary orbits as circles, or even ellipses, which wrap around a star. The planet then travels along this fixed path, over and over, repeating its exact positions again and again.

Most of us imagine this incorrectly. As it turns out, these orbits precess, meaning the orbit itself rotates around a star. In other words, if you were to take a piece of paper and plot the orbit over time, instead of going over the same line over and over again, it would shift ever so slightly, and produce more of a "floral" pattern as the planet's orbit itself revolves around the Sun. Newtonian mechanics predict such a shift. In the case of Mercury, however, the shifting and precession predicted by theory were not enough to explain the degree of shift found by observation. The orbit spins more rapidly than anyone had predicted.

1.3 Speed of Earth and the Ether

In the physics of the 19th century, it was thought that light was a wave with no particle properties at all, and that it couldn't travel without some sort of material or medium to travel through. As it was clear that light reached Earth from the Sun, and as the finite extent of the atmosphere had been established, that meant there was something out there in space that wasn't air between the Earth and the Sun. Thus was born an idea in physics that made so much sense at the time that some people still won't let it go today: the ether.

The ether was a material thought to be of exceptionally low density that spread throughout the heavens, and it was this ether that light travelled through. As it seemed to have no measurable impact on orbital mechanics, it was also suggested that all of the ether moved or flowed at a steady speed, and that planets sailed through it like a gentle ocean current. This begged an obvious question: how fast is the current?

The speed of the current could be measured with a relatively simple apparatus. Light waves have peaks and valleys, technically called crests and troughs, just as water waves do. As with water, when crest meets crest, they combine to form a higher crest. Similarly, two troughs combine to form a deeper trough. This is called constructive interference. When two waves are "in phase," it means their crests and troughs always combine in this manner. When the crest of one wave combines with the trough from another, they cancel each other out through a process known as destructive interference, and are said to be "out of phase."

This principle was applied to measure the speed of the Earth relative to the ether. Imagine two swimmers in a lake, starting at a floating platform. Now imagine that they are in a race, and have to swim from the floating platform to a landmark in the lake and then back to the platform. Further imagine that each swimmer has his or her own landmark to swim to, and that the landmarks are the same distance from the platform, but that they are in different directions, separated by a right angle. If the swimmers swim at the same speed, they would be expected to swim to the landmarks and back in the same amount of time. If the swimmers were waves, they'd arrive "in phase."

Now take this same setup with our swimmers and put them in a river. Even travelling at the same speed, if one swimmer swims with and against the current, and the other swims at a right angle to the current, they would not be expected to return to the floating platform (which also drifts with the current) at the same time. This is because the flow of the current would ensure that the swimmers were moving at different speeds. If the swimmers were waves, they'd arrive "out of phase."

Albert Michelson and Edward Morley decided to use this principle to measure the speed of Earth relative to the ether. They set up an apparatus using one spinning mirror and two stationary mirrors to split a single beam of light into two beams that travelled at right angles to each other before reforming. They could then measure how "out of phase" the light beams were upon their return and recombination, and determine the speed of the apparatus with respect to the ether along a particular direction at that time. By repeating the experiment several times in the course of a year, the Earth's motion relative to the Sun could be measured and eliminated, and the motion of our solar system relative to the ether could be quantified.

The problem they had was this; no matter what time of year they did the experiment or what size apparatus they used, the light returned in phase every time without exception. For some reason, the speed of light wasn't changing, with or without an ether. It appeared that the entire theory of the ether was fundamentally flawed and that there was no ether. That's not the conclusion most scientists reached, however. The majority of the thinking was not in looking for an alternative to ether, but rather in looking for a flaw in the Michelson- Morley experimental design, despite the number of independent confirmations that were coming in from those who reproduced the experiment. Others looked at creating a more complicated picture of the ether which would have the same result.

1.3.1 The Ether Today

Every reproducible experiment conducted over the past 125 years indicates that the ether does not exist. Despite this complete lack of supporting evidence, there are still those who cling to the theory. Some seem to do so for the novelty, as it has almost become a game to discern how complicated ether must be in order to remain consistent with all of these experiments. Others do so for other reasons. For example, ether is a fundamental piece of most theories which attempt to include existence of ghosts and other paranormal activities in modern science.

In short, people still talk about theories involving ether despite a complete lack of supporting evidence. If you encounter a theory involving ether, take it not with a grain of salt, but with an industrial size salt lick.

1.4 Compatibility of Newton and Maxwell

In the late nineteenth century, the world of physics' most legendary figure was Isaac Newton. Newton's laws of motion had been formulated in the 17th century, and stood unchallenged over 200 years later, having ushered in a new era in predictive, quantitative science. They agreed with experiments to an amazing degree.

In the latter half of the 19th century, the theories of electricity and magnetism started to mature and formal structures started to appear. Through a combination of deriving some relationships and compiling the work of others, James Clerk Maxwell assembled the so-called "Maxwell's Equations," which described electricity and magnetism effectively.

Now, Maxwell's equations had been amazingly successful. Not only did they accurately describe the behaviours of anything involving both electric charges and magnetic fields, they also allowed Maxwell to make the first (reliable) prediction of the speed of light. Light is a combination of electric and magnetic fields that feed off of and propagate each other. Using only some physical constants related to these waves, Maxwell calculated the speed of light.[1] Before that time, the speed of light was a quantity known only by experimental measurements, and had a considerable margin for error. Maxwell's calculation came very close to the now accepted value for the speed of light of $c = 299792458 \ m/s$. Not only was it the first theoretical

---

[1]To be pedantic, as any good mathematician or theoretical physicist must be, Maxwell actually calculated the speeds of electric and magnetic fields independently and then proved they were not only identical, but that the two calculations were closely related. It was this consistency in both result and procedure that helped cement his findings in the minds of his peers.

value for the speed of light, it was a value high enough to explain why the experimentalists had a hard time computing the value correctly. In these early days, the work of Newton and Maxwell appeared to join together like a marriage made in heaven. In spite of these triumphs, there was a serious problem. To illustrate this problem, we must first discuss reference frames.

1.4.1   Reference Frames

In science, all experiments require a well defined *reference frame*.[2] It is, basically, a set of standard positions and directions that are used to measure things against.

For example, imagine you are doing the first physics experiment in a high school physics course. This usually involves dropping something from a given height, and measuring how much time it takes to fall.[3] The distance fallen is typically measured relative to the ground, and the analysis of the accelerations and speeds involved are all relative to the laboratory bench and floor.

This typical and effective setup completely ignores the fact that the Earth rotates during the day about its axis, and the entire planet is revolving around the Sun. The entire solar system also orbits around the centre of the galaxy, and our galaxy may be in motion with respect to some other as yet undetermined body. These motions can be safely ignored in the high school laboratory.[4] We have chosen that laboratory as our frame of reference; we measure with respect to a point in the lab, and can continue unrestrained from that perspective. This is a valid assumption, provided that all frames of reference are equivalent from a scientific standpoint.

Note that we use the word "equivalent" instead of "equal" here, and with good reason. If two reference frames are equal, the quantities measured in one will be an exact match to quantities measured in the other. If they are equivalent, then observers in the two frames would be able to accurately predict future events, even though they measure different numbers for the quantities involved. For example, an observer in the high school laboratory might say that an object has zero speed, velocity, and energy of motion before it is dropped, while an observer on the Moon (which is moving relative to the Earth's surface) would disagree. Both observers would be able to accurately predict what would happen when the object is dropped, so the frames are equivalent, but not equal. Specifically, they both predict that the falling object would take the most direct path possible when moving towards a collision with the laboratory floor.

In Newton's world, these frames are equivalent. When Maxwell's equations are

---

[2]The term "reference frame" is interchangeable with "frame of reference." The latter is somewhat old fashioned, and is not seen as often these days.

[3]Physics experiments at the high school level can be painfully boring. I can assure you, they do get better. Eventually. We need to start students with the most basic experiments to make sure they have their experimentation skills down well enough to move on to the exciting, interesting experiments, which may or may not involve violent explosions and supersonic projectiles.

[4]The Earth's rotation does have an impact measurable by sufficiently advanced equipment, but we'll leave that discussion for later. If it bothers you now to ignore this piece, rest assured that the Earth's motion could be completely ignored if it were linear instead of rotating, and that high school equipment budgets make it extremely difficult to design high school experiments capable of detecting the difference.

combined with Newton's theories, we find that two different frames of reference are not equivalent.

1.4.2 How Charged Objects Interact

Experiments had confirmed a number of behaviours of electrically charged objects. Maxwell's equations formalized these behaviours, but that's all they did; interactions between magnetic fields and electrically charged objects were jammed into the theory because they were observed in the lab, not because they were predicted by existing theory.

When charged objects are stationary, the basic phenomena are explained in a very simple fashion: unlike charges attract, while like charges repel. If you arrange a number of electrically charged marbles on an insulated surface and prevent them from moving, Maxwell's equations will accurately predict the attractive and repulsive forces they all experience.

When charges move, things get complicated. Moving charges produce magnetic fields[5] which can also attract or repel. These behaviours depend on the directions the charged objects are moving as well the charges themselves. Specifically:

- Like charges moving in the same direction produce attractive magnetic fields.

- Like charges moving in opposite directions produce repulsive magnetic fields.

- Unlike charges moving in opposite directions produce attractive magnetic fields.

- Unlike charges moving in the same direction produce repulsive magnetic fields.

The faster the charges move, the stronger the magnetic fields they produce, and the more pronounced the forces become. The same is not true of the electric forces: they are entirely independent of the relative motions of the particles.

1.4.3 Newton to Maxwell: "It's Not Me, It's You"

Imagine two wires, running parallel to each other for as far as the eye can see. In fact, imagine them to be running farther than the eye can see; imagine they are infinitely long. Now, imagine that they carry identical electrical charges, and that these charges are spread uniformly across the wires. No one point on either wire is distinguishable from any other point on the same wire.

An observer standing next to the stationary wires would predict a repulsive force, and the two wires would move apart. That is the result of the laws of physics as measured in his or her reference frame.

Now imagine a second observer flying along parallel to the wires, exactly half way between them. From this observer's reference frame,[6] the wires are in motion, and

---

[5]Why do moving charges produce magnetic fields? In Maxwell's time: "because they do." This was discovered in the lab, and subsequently jammed into the theory because it needed to be there, and not because any theorist could predict or explain the phenomenon.

[6]An observer's reference frame is one centred on that observer in which the observer is always treated as stationary.

moving in the same direction, so the wires experience *two* forces. In addition to the repulsive electrical force, there is also an *attractive* magnetic force. The electric force is independent of the relative motions of the wires. The magnetic force, however, increases with speed. As our flying friend accelerates, the attractive magnetic force gets stronger and stronger, until it eventually overpowers the repulsive force and the wires approach each other instead.

This is a significant discrepancy. Do the wires move apart, as predicted by our standing observer, or do they collide, as predicted by our flying observer? These predictions are not equivalent, and so neither are the two frames of reference.

Newton's theories were those that not only governed motion, but which also described how to work in moving frames of reference. His theories had stood up to every conceivable experimental test for more than 250 years. As a physicist, he was revered by those working in the fields of research, and seemed more than human through the lens of history.[7] Maxwell, on the other hand, was very human, and his equations were relatively recent discoveries, having been collected and studied for less than 20 years. Thus, the overwhelming majority of physicists turned to Maxwell, looking for his mistake, and thereby fixing the problem with incompatibility.

Now, physicists are human, and when a theory that stands as long as Newton's is shown to be incompatible with something new and exciting, there are two natural reactions. The first, and by far most common, is to assume that the older Newtonian theories are correct and the new ones are wrong. However, in the world of faceless science, there are young upstarts excited about the prospect of becoming famous by rewriting the "old and busted" theory to force it to conform to the "new hotness" theory. One such upstart was Hendrik Antoon Lorentz, who, at the tender age of 51, decided to determine exactly how to rewrite Newton to make his work compatible with Maxwell's. His work resulted in an absolute absurdity: to reconcile the two theories, time had to depend on the speed of the observer! To most, this constituted further proof that the problem lay in Maxwell's work.

Unfortunately, no problem could be found. Many scientists started connecting the problem to theology, deciding that somewhere out there was a reference frame that worked. In other words, they believed that detailed explorations and experimentations would reveal situations that were not consistent phenomenologically with a laboratory based experimental reference frame. By determining which reference frame was consistent with such experiments, they would be able to determine God's preferred frame of reference. They intended to then search the heavens for such a body, in hopes of discovering the location of Heaven. The early results of this search were inconsistent and fruitless.

1.5   Enter Einstein

The world of professional physicists scoured the experimental and theoretical worlds looking for solutions to these problems. It took a self-taught physicist employed as

---

[7]The lens of history is about the only lens Newton didn't completely describe up to that point, which was likely a contributing factor.

a patent clerk to see something nobody else had seen in the work. Most physicists were excited because "Maxwell had calculated the speed of light using only some physical constants!" It was Einstein who got excited that "Maxwell had calculated the speed of light using only some physical constants!" Einstein realized that certain expected information was absent from the calculation. The full implications of this realization and the birth of the special theory of relativity are the subject of the next chapter.

The remainder of this chapter adds mathematical details relevant to the above discussion. Familiarity with vector calculus is assumed.

1.6 Mercury's Orbital Precession

The full calculations of Mercury's precession will be omitted. The basic argument is this: Mercury rotates on its axis as it orbits around the Sun, as do the other planets in our solar system. Thus, it carries angular momentum. This angular momentum experiences a torque as the gravity of the other planets in the solar system and the Sun play off of each other. This causes all planets, including Mercury, to shift their orbits. The orbits rotate around the Sun in the direction of the Sun's rotation, and the magnitude of this orbital rotation ("precession") depends on the gravitational interactions of all local bodies and the eccentricity of the precessing orbit. Quantitatively, Mercury's orbit was precessing about 41" more than it should have each century.[8]

1.7 Calculating the Speed of Light

In Maxwell's day, the general form of the wave equation was well known. If a wave described by function f propagated with speed v, then it satisfied the wave equation

$$\nabla^2 f = \frac{1}{v^2} \frac{\partial^2 f}{\partial t^2}$$

where

$$\nabla = \begin{pmatrix} \frac{\partial}{\partial x} \\ \frac{\partial}{\partial y} \\ \frac{\partial}{\partial z} \end{pmatrix}$$

is the usual vector differential operator.[9]

---

[8]For those unfamiliar with this angular measure, 60″ ("sixty seconds of arc") is equal to 1′ ("one minute of arc"), and in turn, 60′ is equal to 1°. In other words, Mercury's orbit was wrong by $\frac{41}{3600}$° per century and the astronomers could measure accurately enough to know that was a problem. Observational astronomers have a history of being very, very good at their jobs.

[9]Granted, in Maxwell's day, vectors were not common mathematical tools in physics. Instead, each vector component had its own equation, which is why some of the variables chosen for electromagnetic quantities are so odd. Maxwell denoted the three components of the magnetic field as A, B and C, and the electric field was D, E and F, and so forth. When the usefulness of vectors was recognized, it was noticed that, by complete and total coincidence, the y component of the electric field was E, the y component of the magnetization (not the magnetic field) was M and the y component of the polarization was P. To make the transition to vectors as easy

In general, Maxwell's equations appear as follows:

$$\nabla \cdot \mathbf{E} = \frac{1}{\epsilon_0} \rho$$

$$\nabla \times \mathbf{E} = -\frac{\partial \mathbf{B}}{\partial t}$$

$$\nabla \cdot \mathbf{B} = 0$$

$$\nabla \times \mathbf{B} = \mu_0 \mathbf{J} + \mu_0 \epsilon_0 \frac{\partial \mathbf{E}}{\partial t}$$

where ρ is electric charge density, $\mathbf{J}$ is the electric current vector, ϵ_0 is the electric permitivity of vacuum and μ_0 is the magnetic permeability of free space. If one is working in a vacuum with no excess charge or current, then $\rho = 0$ and $\mathbf{J} = \mathbf{0}$, so that the equations reduce to

$$\nabla \cdot \mathbf{E} = 0$$

$$\nabla \times \mathbf{E} = -\frac{\partial \mathbf{B}}{\partial t}$$

$$\nabla \cdot \mathbf{B} = 0$$

$$\nabla \times \mathbf{B} = \mu_0 \epsilon_0 \frac{\partial \mathbf{E}}{\partial t}$$

Maxwell noticed that the two fields can be decoupled, or separated into equations with only $\mathbf{E}$ or $\mathbf{B}$ and not both, through use of the identity

$$\nabla \times (\nabla \times \mathbf{A}) = \nabla (\nabla \cdot \mathbf{A}) - \nabla^2 \mathbf{A}$$

for any possible $\mathbf{A}$. For example, starting with

$$\nabla \times \mathbf{E} = -\frac{\partial \mathbf{B}}{\partial t}$$

and "taking the curl" by calculating the cross product with ∇ positioned on the left gives

$$\nabla \times (\nabla \times \mathbf{E}) = \nabla \times \left(-\frac{\partial \mathbf{B}}{\partial t} \right)$$

Working with the left hand side first, recognizing that $\nabla \cdot \mathbf{E} = 0$, we get

$$\nabla \times (\nabla \times \mathbf{E}) = \nabla (\nabla \cdot \mathbf{E}) - \nabla^2 \mathbf{E} = -\nabla^2 \mathbf{E}$$

Meanwhile, noting that ∇ and $\frac{\partial}{\partial t}$ are operators that commute with each other, we

---

as possible for those comfortable with the old notation, it was decided that the symbol for the y component of every vector would become the symbol for the entire vector, since that allowed for three intuitive labels, which was typically three more than any other system that was compatible with both notations.

find that the right hand side becomes

$$\nabla \times \left(-\frac{\partial \mathbf{B}}{\partial t} \right) = -\frac{\partial}{\partial t} \left(\nabla \times \mathbf{B} \right)$$

$$= -\frac{\partial}{\partial t} \left(\mu_0 \epsilon_0 \frac{\partial \mathbf{E}}{\partial t} \right)$$

$$= -\mu_0 \epsilon_0 \frac{\partial^2 \mathbf{E}}{\partial t^2}$$

Thus,

$$-\nabla^2 \mathbf{E} = -\mu_0 \epsilon_0 \frac{\partial^2 \mathbf{E}}{\partial t^2}$$

or

$$\nabla^2 \mathbf{E} = \mu_0 \epsilon_0 \frac{\partial^2 \mathbf{E}}{\partial t^2}$$

This fits the general form of the wave equation perfectly, provided

$$\frac{1}{v^2} = \mu_0 \epsilon_0$$

or

$$v = \frac{1}{\sqrt{\mu_0 \epsilon_0}}$$

Therefore, it appeared that the electric field would propagate through vacuum with a speed $v = \frac{1}{\sqrt{\mu_0 \epsilon_0}}$. A similar calculation can be performed for $\mathbf{B}$ to show that field also propagates with this speed. Maxwell performed this calculation, and found that this was a perfect fit for the experimentally determined speed of light! This was how physicists first realized the nature of light was that of an electromagnetic wave.

1.8 The Michelson-Morley Experiment

The Michelson-Morley experiment comes down to the difference in transit time for the light going through two paths. We start by imagining our apparatus is moving with one arm parallel to the motion of the ether, and the other arm perpendicular to the motion of the ether:

The light that travels parallel to the ether will move at two different speeds. With the speed of light c and the speed of the ether v_e, it will move with total speed $c + v_e$ when it moves with the current of the ether and with total speed $c - v_e$ when it moves against the current. Thus, the transit time there and back for the light moving parallel to the ether (given an arm length d in the apparatus) is

$$t_\parallel = \frac{d}{c + v_e} + \frac{d}{c - v_e} = \frac{2dc}{c^2 - v_e^2}$$

The light that moves perpendicular to the ether, on the other hand, takes the same amount of time to travel each trip, and is not affected by the speed of the ether. As a result, the time for that light to make the round trip is

$$t_\perp = \frac{2d}{c}$$

The difference in times can be calculated as well:

$$t_\parallel - t_\perp = \Delta t = \frac{2dc}{c^2 - v_e^2} - \frac{2d}{c} \tag{1.1}$$

With sufficient algebraic manipulation, we find that

$$v_e = \pm\sqrt{1 - \frac{2d}{c\Delta t + 2d}} \tag{1.2}$$

The difficult part is pinning down Δt. This is an incredibly small time difference to measure. Thankfully, none of this motion is expected to alter the frequency of the light in question. Thus, the period of the light is the same. If the period of the light is $1.50 \times 10^{-15}s$, as it would be for the shortest wavelength of light the average human eye can detect, then counting the difference in periods can be extremely difficult. If the period of the light is T, then we can break the time interval Δt into the number of periods that have elapsed n with the simple formula

$$n = \frac{\Delta t}{T}$$

Unfortunately, since Δt is being measured by a wave interference pattern, there is a difficulty. The interference pattern is not unique for a particular n. In terms of the mixed numbers we all saw in our early exposure to fractions, the whole number portions are ignored. Thus,

$$n = 1\frac{1}{3} = 2\frac{1}{3} = 3\frac{1}{3} = 4\frac{1}{3} = 5\frac{1}{3} = 6\frac{1}{3}$$

are all indistinguishable cases. This is why the experiment takes so long to do; measurements must be taken year round as Earth's annual orbit around the Sun change its direction of motion with respect to the ether, so that sufficient data can be collected to determine precisely which value of n is the correct one to use.[10] In the end, the difficulties were somewhat moot, as every attempt to collect data from the experiment was consistent with $n = 0$.

1.9   Force Between Two Wires

Imagine the two wires spoken of before, parallel and separated by distance d. When they are stationary and carrying electrical charge, each with charge density λ, the electric force per unit length they experience is given by

$$\frac{F_E}{l} = \frac{\lambda^2}{2\pi\epsilon_0 d}$$

which is a repulsive force. When they are stationary, there is no magnetic field. From the perspective of a moving observer, however, they do produce magnetic fields, and the magnetic force per unit length they experience is the attractive force

$$\frac{F_B}{l} = \frac{\mu_0 \lambda^2 v^2}{2\pi d}$$

---

[10]The exact calculations Michelson and Morley did differed from this in details, but this illustrates the concept effectively and is a bit easier to grasp when one's skills are rusty.

The issue comes in because of the difference in signs. The force between the two is completely balanced when

$$\frac{\lambda^2}{2\pi\epsilon_0 d} = \frac{\mu_0\lambda^2 v^2}{2\pi d}$$

or

$$v^2 = \frac{1}{\epsilon_0\mu_0}$$

So, any observer travelling at $v < \sqrt{\frac{1}{\epsilon_0\mu_0}}$ would observe a repulsive force, while any observer travelling at $v > \sqrt{\frac{1}{\epsilon_0\mu_0}}$ would observe an attractive force. Those travelling at $v = \sqrt{\frac{1}{\epsilon_0\mu_0}}$ would observe no net force at all.

Chapter 2

The Revelations of Einstein

2.1 Einstein's Revelation

In the early 1900s, physicists were confused. Electricity and magnetism, as described by Maxwell's equations, seemed incompatible with Newton's work in motion. The incompatibility arose when two different observers studied the same phenomenon from two different frames of reference. On top of that, they had a hard time studying the ether they had proposed as a fundamental part of their theory of light, as they could not produce the interference effects one would expect when the speed of light changed with respect to the medium the light was travelling through.

In 1905, Einstein solved both of these problems with one revelation.[1] Scientists were thrilled with Maxwell's calculation of the speed of light, which was derived solely from the values of a few physical constants. Einstein was the first to recognize that a fundamental piece of information appeared to be missing from the calculation. Maxwell's calculation of the speed of light was completely unrelated to the observer's frame of reference.

In other words, two different observers looking at the same light wave would agree on the speed the light was moving, regardless of how they were moving with respect to each other. They do not necessarily agree on the velocity, as that includes direction, but they will agree on the speed.

The manner in which this explains the Michelson-Morley result is relatively clear: if you cannot produce a difference in the speed of light along the two paths, you cannot produce different interference patterns by moving the apparatus. The manner in which this reconciles with Maxwell is far more involved; we will need to lay some groundwork before we can cover that.

---

[1]The third problem mentioned in the previous chapter, dealing with Mercury's orbit, would also be solved by Einstein, but not for another ten years.

2.2   Postulates

In mathematics, when one is formulating a new theory or idea, one begins with "axioms," or ideas that are assumed to be true. The equivalent statements in the experimental sciences are called "postulates." A postulate is something that seems reasonable and is assumed to be true based on daily experiences or scientific observations. A properly formed physics theory begins with postulates that are thoughts formed in some spoken language, which is then translated into mathematics and the implications are followed.

2.2.1   Newton's Postulates

When Newton did his work in motion, he proposed several postulates, including three laws and six corollaries. He also made a few assumptions that he did not formally propose as a part of his system. The laws and his key assumptions are listed here:

1. Law 1: Unless an outside force is applied, a stationary object (object "at rest") will remain stationary and a moving object will continue moving at the same speed and in the same direction.[2]

2. Law 2: If one does apply a force to an object, that object will accelerate in the direction the force is applied. Greater forces create greater changes in speeds. These changes are proportionate to both the force and the "quantity of matter" in an object. Newton originally proposed that the "quantity of matter" was the inertia of an object, and then suggested that the mass of an object is that inertia, so that mass and inertia were identical, which is why inertia is discussed conceptually in class and then mass is used in the algebra.

3. Law 3: If body A pushes against body B with a given force in a given direction, body A also feels a force of equal magnitude and opposite direction coming from body B.

4. Unstated assumption 1: In all of Newton's work, though he didn't formally state it, he assumed that two different individuals measuring the elapsed time between two events would always measure the same interval of time. In other words, if some poor kid's scoop of ice cream falls off its cone at a fair, the time it takes that ice cream scoop to fall to the ground would be the same amount of time whether measured by that child's parent or by someone riding the ferris wheel.

5. Unstated assumption 2: This, technically, was unstated in the *Principia Mathematica*[2] in which he detailed his laws of motion, but it was written in New-

---

[2]Prior to Newton, it was believed amongst the Christian cultures that moving objects required continuous outside forces to stay in motion. Objects moving along the ground stopped not due to friction but due to the lack of a continuous outside driving force. The stars, planets, moons and other heavenly bodies remained in motion because, it was believed, God had tasked angels with the job of pushing them constantly around in the sky for the amusement of humans. The author is unaware of any suggested reason that God disliked these angels so much that He doomed them so such a monotonous eternity, nor is he aware of alternate theories proposed by the non-Christian cultures.

ton's work in theology. Newton recognized the possibility of nonequivalent reference frames, and was one of the first to propose that a "preferred" reference frame could be found that would indicate the location of the Christian Heaven.

The corollaries are clarifications of how the math behind the three laws works, and will not be discussed here in detail.

2.2.2 Einstein's Postulates

Einstein proposed two postulates in 1905 that turned the physics world on its head, eventually. The initial reaction to Einstein's crazy ideas was utter disbelief in response to some of the mathematical implications.

1. All "inertial" reference frames are valid and equivalent. An "inertial" reference frame is one in which the origin of the coordinate system is not being accelerated. Thus, you can measure relative to a stationary object or to an object moving at a constant speed, but not relative to someone riding a ferris wheel, since that person's direction is changing. (Direction counts for acceleration.) In 1915, Einstein would formulate a new theory that would allow the use of any reference frame, treating them all as equivalent, but that addition will come in a later chapter.

2. The speed of light is "invariant," meaning all observers in all valid reference frames measure exactly the same value for the speed light is moving at.

Einstein felt he was on the right track when he used these postulates to derive new ways to adjust the math for changing reference frames, and arrived at a result *identical* to the one that H. A. Lorentz found when he tried to reconcile Maxwell's equations with Newton's laws. That seemed to be a very compelling coincidence.

The complete implications of these postulates are not obvious, particularly if one is trying to see the connection between these ideas and the famous $E = mc^2$ equation associated with relativity. We will examine some of the implications of these postulates from now until the end of chapter six. In chapter seven, we will add the postulates that allow reference frames to experience acceleration, and deal with those implications for the rest of this book. These early chapters deal with the special theory of relativity, since they relate to the "special case" in which gravity and other outside fields and forces do not apply. The latter chapters deal with the general theory of relativity, in which gravity and accelerations are included, so that we can work with "general' cases of physical situations.

2.3 Implications for Distances

Let us begin by examining the way distances are measured by moving observers under Newtonian mechanics and reference frames.

Imagine we have two observers, named Hal and Barry. Hal is standing still, while Barry is running at a constant speed. Both are observing a car driving in the same direction Barry is running and at a higher speed than Barry is running with. When

Hal measures the distance the car has travelled, he measures relative to himself. Barry also measures the distance the car has travelled relative to himself, but he is between Hal and the car, so he concludes that the distance travelled by the car is *less* than the distance Hal measures.

"But wait!" one might cry. "Why not measure relative to the road to eliminate this problem?" This is what is commonly done in labs for consistency. However, the important thing to recognize is that there is *no* problem here. If all inertial frames are equivalent, then it doesn't matter *who* does the measurement, provided the frame of reference is described accurately in the results. This helps in cases such as the motion of astronomical bodies, in which case there is no clear "road" to use as a point of reference.

On a conceptual level, there is no difference between the Newtonian view and the Einsteinian view. The observer moving in the same direction as the car will measure a smaller distance travelled than the observer who is stationary relative to the road. Once the math is examined in detail, one finds that the difference between the two measurements is more dramatic in the Einsteinian view, but the basic concept is the same. The difference is due to the invariance of the speed of light; there are no "instantaneous" measurements, since information travels no faster than the speed of light.[3] In short, Hal and Barry don't measure the position of the car "now," but instead measure the position of the car according to where it was when it reflected the light their eyes are detecting "now." This lag causes the difference between the Newtonian and Einsteinian measurements observed here.

This has further implications. A car has a volume; what if they measure the distances to both the front and rear of the car? The time lag due to the transit time of light is more pronounced from the front (further) side of the vehicle than it is from the rear (closer) side. Thus, if you ask both Hal and Barry to measure the *length* of the car, they would provide two different answers! In relativity, objects moving relative to the observer contract or shrink, and do so in greater degree as the differences in speed become more pronounced. This is known as *length contraction*.

2.4   Implications for Time Intervals

Under Newtonian mechanics, Hal and Barry would disagree about the distance the car travelled in our previous example, but they *would* agree on the amount of time the car spent in motion. The invariance of the speed of light throws a monkey wrench into that perspective.

Imagine the same situation as above. This time, however, the car's headlights are on. Instead of measuring the distance and time elapsed of the car's travel, Hal and Barry measure those quantities for the light waves emitted by the car's headlights.

Just as before, they measure the distance travelled by that light, and come up with different results. In Newtonian mechanics, they would each measure the speed of

---

[3]These chapters do not deal with some quantum mechanical phenomena such as entanglement or recent neutrino results which may indicate exceptions to this rule. Look for separate chapters on those topics when they are better understood.

that light by measuring the elapsed time, and then dividing the distance travelled by the elapsed time. As they would agree on that time interval, they would arrive at two different numbers for the speed of light. In Einstein's relativity, this is impossible; all observers measure light in a vacuum to be travelling at the same, invariant speed. What implications does that have?

We can flip the math around. Both observers measure the distance travelled, though they disagree on the exact value of this distance. They also measure the speed of light, and both arrive at the same value of $c = 299792458 \ m/s$. The time elapsed can be calculated by taking the distance travelled and dividing by the speed of light. Hal and Barry each do this calculation, using *different* numbers for the distance travelled, but the same numbers for the speed of light: they arrive at *different* numbers for the time elapsed, each of which is correct in that observer's reference frame!

This was a staggering thought. Time depends upon the speed of the observer. This phenomenon, known as *time dilation*, has since been verified by many, many experiments. This is a real and undeniable effect. If someone observes a moving clock, he or she would conclude that the clock in motion is running more slowly than an accurate clock which is considered to be "at rest."

2.5   Implications for Velocities

Let us return to Hal, Barry and the car. This time, instead of running in the same direction as the car, Barry runs in the opposite direction. Let us also assume that, as Hal measures things, both Barry and the car are moving at more than half the speed of light; for the sake of argument, let us choose three quarters of the speed of light as the speed they travel.

In the Newtonian view, if Barry were to measure the speed of the car, he would calculate the speed that is simply the sum of his speed and the car's speed, as measured by Hal. In this case, that would result in an answer that is one and a half times the speed of light. In the Einsteinian view, this changes dramatically.

For the sake of argument, let us assume the car's headlights are still on. When Barry measures the speed of the light coming from those headlights, he concludes that it travels at the speed of light $c = 299792458 \ m/s$, as he must. When he measures the speed of the car itself, he finds that it is *slower* than the speed of light; the light wave from the headlights are further away from the car, and getting further away all the time.

In other words, the Einsteinian view allows for agreement about *relative* speeds of bodies, but not *absolute* speeds. Simple comparisons such as "faster" and "slower" are maintained, but the actual numbers vary considerably. This is because the observers don't just disagree on the distances travelled, but also on the time elapsed. The truly stunning implication is this: *nothing* can travel faster than the speed of light. Any moving object can "turn on its headlights" or not, and it will not affect how others measure the speed of that body. Everybody will, however, agree that the hypothetical light from those headlights travels faster than the source (the moving object) and that the light is travelling at the speed of light. Therefore, all observers agree that the source of the light is moving slower than light.

2.6   Einstein and Maxwell

Now that these simple implications have been discussed, we can return to our paradox from electricity and magnetism. Recall the situation with two infinitely long and identically electrically charged rods. From a reference frame in which they are stationary, the rods will repel each other due to the like charges on the rods. The more dense the electrical charges are on those rods (i.e. the more charge is present in any given length of the rod) the stronger this repulsion becomes. From a moving reference frame, these charges form a current, which results in a magnetic attraction. If one moves quickly enough, the laws of nature described by Newton and Maxwell would predict that they would eventually attract each other more strongly than they would repel, creating a paradox: the rods cannot repel from Hal's perspective but collide from Barry's perspective.

The Einsteinian model resolves this paradox in a somewhat surprising way. The charge density on the two rods plays just as strong a role in generating the electrical force as it does the generating the magnetic force. As a result, every situation changes from a repulsive force to an attractive force at the exact moment the moving observer reaches the speed of light, which is not possible. (If Barry has a flashlight in hand, he will never outrun the light that comes out of it, so he must always be travelling at less than the speed of light.) True, Barry will observe a stronger force than Hal observes, but he also measures less elapsed time than Hal, so the overall effect results in the same predicted motion of the two rods. This completes the conceptual content for this chapter. Now, we will see the math that backs this up.

2.7   The Lorentz Transformations

The method of derivation used by Albert Einstein[1] will be followed here. That used by Lorentz is significantly longer and will be omitted.

We begin by examining a photon in two different reference frames, which we label S and S'. Each frame contains an observer who considers himself or herself to be stationary, or at rest, with respect to the other frame. The origins of the two frames match up perfectly, so that when $x = y = z = ct = 0$ then $x' = y' = z' = ct' = 0$, and the axes align. (i.e. at time $t = t' = 0$, x is parallel to x' through the common origin, y is parallel to y' through the common origin, and z is parallel to z' through the common origin.) From the perspective of an observer in frame S, the frame S' moves in the positive x direction with speed v.

Assume the photon being observed passes through both origins and is also travelling in the positive x (and positive x') direction. By Einstein's postulate, it travels with speed c in both reference frames, so it satisfies $c = \frac{x}{t} = \frac{x'}{t'}$ automatically. It will be convenient to rearrange these formulae as $x - ct = 0$ and $x' - ct' = 0$ instead. By logic, we assume that there is a function λ which depends solely on v such that the

relationship
$$(x - ct) = \lambda \left(x' - ct' \right)$$

is satisfied. From the S' frame, it can be argued that frame S is moving in the negative x' (x) direction with speed $-v$, so we also have the similar relation

$$\left(x' + ct' \right) = \mu \left(x + ct \right)$$

Adding and subtracting the two transforms them into the relations

$$x' = \gamma x - \delta ct \tag{2.1}$$

and

$$ct' = \gamma ct - \delta x \tag{2.2}$$

where

$$\gamma = \frac{\lambda + \mu}{2}$$

and

$$\delta = \frac{\lambda - \mu}{2}$$

If we observe the origin of S' now, instead of the photon, we have $x' = 0$ at all times. We can rearrange equation (2.1) to obtain

$$\frac{x}{t} = v = \frac{\delta c}{\gamma}$$

Now, let us combine equation (2.1) and equation (2.2) by first isolating t in equation (2.1):

$$t = \frac{\gamma x - x'}{\delta c}$$

and substituting it into equation (2.2):

$$ct' = \frac{x}{\delta} \left(\gamma^2 - \delta^2 \right) - \frac{\gamma}{\delta} x'$$

This general expression should hold for any object, including one travelling along the x (and x') axis in such a way that one end passes through the origin of both coordinate systems while the other end is on the positive x' (and x) axis. If we measure the length of this object at this moment, then $t' = 0$ and we have

$$x' = \gamma x \left(1 - \frac{v^2}{c^2} \right)$$

We can apply the same logic in either reference frame, and arrive at the overall formula

$$\gamma^2 = \frac{1}{1 - \frac{v^2}{c^2}}$$

or (selecting the positive root)

$$\gamma = \frac{1}{\sqrt{1 - \frac{v^2}{c^2}}} \qquad (2.3)$$

Complete substitution results in the Lorentz transformations:

$$x' = \frac{x - vt}{\sqrt{1 - \frac{v^2}{c^2}}}$$

$$y' = y$$

$$z' = z$$

$$t' = \frac{t - \frac{v}{c^2}x}{\sqrt{1 - \frac{v^2}{c^2}}}$$

It is common to use equation (2.3) to express these as

$$x' = \gamma (x - vt) \qquad (2.4)$$
$$y' = y \qquad (2.5)$$
$$z' = z \qquad (2.6)$$
$$t' = \gamma \left(t - \frac{v}{c^2}x \right) \qquad (2.7)$$

and we will do so for most of this book.

We can also examine frame S from the frame S' and find the similar equations

$$x = \gamma (x' + vt') \qquad (2.8)$$
$$y = y' \qquad (2.9)$$
$$z = z' \qquad (2.10)$$
$$t = \gamma \left(t' + \frac{v}{c^2}x' \right) \qquad (2.11)$$

2.8   Length Contraction and Time Dilation

Once the Lorentz transformations have been established, it becomes a straightforward matter to quantify the length contraction. We begin with equation (2.4):

$$x' = \gamma (x - vt)$$

Now, if we wish to determine the length of an object moving in frame S' as measured by an observer in frame S, we measure the positions of the two ends of the object and subtract them so that $l = x_2 - x_1$. We have the freedom to choose our origins of the coordinate systems as before, with the "trailing" edge of the moving object passing through the origins of both coordinate systems. With that definition, $x_1 = 0$

and we can drop the subscripts. We also measure both ends simultaneously[4] as the trailing edge passes through the origin, so that $t = 0$. This planned definition of variables makes the calculation relatively simple:

$$x' = \gamma x$$

Remembering that the object is at rest in frame S', we find that

$$l = l'\sqrt{1 - \frac{v^2}{c^2}}$$

When $v < c$, we find that $l < l'$ (as $\sqrt{1 - \frac{v^2}{c^2}} < 1$), and the length is contracted.

Time dilation is calculated by virtually identical math to length contraction. When an observer in S watches a stopwatch positioned at the origin in frame S', he or she finds that

$$t = t'\sqrt{1 - \frac{v^2}{c^2}}$$

2.9 Velocity Transformations

The velocity transformation is fairly easy from a conceptual standpoint. From a notational standpoint, we already use the symbol v as the speed of the frame S' relative to frame S, so we need a different symbol. Therefore, let the velocity we are concerned with be denoted as

$$\mathbf{u} = \begin{bmatrix} u_x \\ u_y \\ u_z \end{bmatrix}$$

or

$$\mathbf{u} = \frac{d\mathbf{x}}{dt}$$

Thus, to calculate the velocity as measured in frame S', we compute

$$\mathbf{u}' = \frac{d\mathbf{x}'}{dt'}$$

In the case when the reference frames move with speed v along the x/x' axes only, then we have $u_y' = u_y$ and $u_z' = u_z$. To calculate u_x', we use the total differentials. We treat v as a constant, so we have

$$dx' = \gamma\left(dx - v dt\right)$$

and

$$dt' = \gamma\left(dt - \frac{v}{c^2}dx\right)$$

---

[4]More on this word later.

Taking the ratio, we have

$$u'_x = \frac{dx'}{dt'}$$

$$u'_x = \frac{\gamma\,(dx - vdt)}{\gamma\,\left(dt - \frac{v}{c^2}dx\right)}$$

$$u'_x = \frac{\frac{dx}{dt} - v}{1 - \frac{v}{c^2}\frac{dx}{dt}}$$

$$u'_x = \frac{u_x - v}{1 - \frac{u_x v}{c^2}}$$

which is the desired result.

Chapter 3

Space and Time

3.1 Space and Time Before Relativity

Prior to Einstein's work, the importance of both space and time to physics was clearly recognized. They were, however, treated as completely distinct quantities.

On one hand, they had the spatial dimensions. These came in three fundamental directions. The scientist had the ability to choose these directions to some degree. On the surface of the Earth, one could choose North as one direction, East as another, and Up as the third. If one chooses North then one does *not* choose South; mathematically, they describe the same line, since you can "undo" all travel in the direction North by travelling in the direction South. Directions are only "different" if they cannot be completely canceled by motion in the previously defined directions. One could choose Northeast as a direction, since the "East" portion can't be cancelled by North, but choosing directions that aren't at right angles to each other usually makes things more complicated than they need to be when working on the math. The directions are connected through rotations: if you rotate through a right angle, the person who was facing North can now be facing East or Up. It is easy to see how they relate to one another, and that close relationship is one of the reasons we have the ability and inclination to define whatever directions are convenient for us. (For example, when working on a ramp, it is more convenient to pick directions parallel and perpendicular to the surface of the ramp than simply North, East and Up.)

On the other hand, there was (and is) time. No clear connection between space and time was apparent. Sure, they could plot graphs using time as one axis and space on another, but that was a technique used to see how systems evolve, and was not considered an indication of any deeper, fundamental connection between the quantities. Time was absolute and the same (invariant) for all observers, while distances were different for observers in motion (though not as different as they are under Einstein; the distance a car travelled was variable under Newtonian mechanics, but the length of the car was not, since the front and back could be measured "simultaneously.")

3.2   Space and Time After Relativity

With Einstein's relativity, we learn that time is not invariant, and that perceived time intervals depend greatly upon the motion of the observer. Moreover, the manner in which one measures distances depends on time measurements and vice versa. This implies a much stronger connection between the two quantities than was previously believed. This begs a number of questions: if time is that closely linked to space, can it be treated as another direction? If so, why does that direction behave so much differently than space? Why can we not navigate back and forth through time as easily as we do in space? Why can't we simply rotate from a spatial direction to a time direction as easily as we rotate from North to East? These questions opened some very productive floodgates. Physicists were forced to examine their beliefs about geometry, which led to some startling revelations when they explored the abstract discoveries made by mathematicians in the recent centuries.

Mathematicians can always be depended upon to indulge their curiosity, regardless of whether or not their discoveries apply to the "real world." One such avenue was the exploration of the square roots of negative numbers. For centuries, it was believed that negative numbers could have no square roots. In fact, they can, although it was difficult for even mathematicians to accept that fact when they were first proposed. This resistance is what led to the square roots of negative numbers being named "imaginary numbers:" even the mathematicians at the time had a hard time accepting them.[1] Despite their name, imaginary numbers are just as valid as any other numbers. Moreover, mathematicians had found that one can include directions describing imaginary numbers in geometry and arrive at precisely the type of system physicists needed. If time were measured along an imaginary direction, then it would be a direction tightly linked to the spatial directions, and yet would still be distinct and unique in precisely the manner required to satisfy intuition. No simple[2] rotation can point someone in an imaginary direction, yet the direction is an integral part of the geometry of reality. This distinction is enough to explain why we cannot control our motion along the time direction: it is a fundamentally different type of direction. Still, this would have been incredibly difficult to accept if not for the discovery of invariants.

3.3   Invariants

The realization that this mathematical framework described the time direction went hand in hand with another accidental discovery that had yet to be explained, related to the lengths of certain mathematical objects.

In Newtonian terms, moving observers disagreed on the positions of objects, but they always agreed on the lengths of those objects. Relativity threw that convenience out the window, but did so in a way that pulled time in as another direction

---

[1]Most high school students have heard of the real numbers. They were named to contrast the imaginary numbers in an act of professional mockery. "You can study those imaginary numbers if you'd like, but *I* work with *real* numbers."

[2]This word is surprisingly important, as we'll see shortly.

to be considered in the geometry. Physicists began working with various ways to combine the space coordinates with the time coordinate to see if this idea of an "invariant" length could be preserved. They had discovered a means to make this happen that was remarkably counterintuitive.

If an object is turned to lie along one of the directions chosen for our coordinate system, then it is relatively easy to measure its length: find the coordinates of the two ends, and subtract the smaller number from the bigger number. Since the days of Pythagoras, there was a known method for measuring the length of an object that didn't lie along one of these directions: measure the coordinates of the extreme ends of the object along each of the chosen directions, square them, add the squares, and then take the square root of the answer. The calculation may be somewhat cumbersome without a calculator, but it is useful to know how to do this when one cannot simply turn the object or the measuring device to measure something in the simple, intuitive manner.

In trying to do this with relativity, physicists noticed something very strange. Adding in the square of the time coordinate didn't resolve the issue: both time and distance were smaller numbers in a moving reference frame. The strange and startling discovery was that these reductions could cancel each other out if subtracted! In other words, if one added the squares of the lengths of the object along the space dimensions, but then *subtracted* the square of the length along the time direction, the answer one got was the same for *all* observers. Quantities that are the same for all observers are known as *invariant* quantities.

This idea was shocking and counterintuitive, but completely consistent with treating time as an imaginary number. Recall from the previous section that imaginary numbers were discovered as the square roots of negative numbers: if one squares an imaginary number, the answer is negative. Adding a negative number is mathematically identical to subtracting a positive number. The implications of relativity get crazier and crazier, and yet somehow remain entirely consistent with each other and with experimental results. Slowly but surely, Einstein's seemingly insane notions were gaining more and more support.

4 Minkowski

Hermann Minkowski laid much of the foundation for this combination of space and time when he addressed the attendees of the 80th Assembly of German Natural Scientists and Physicians[3] on September 21, 1908.[4] His speech began as follows (after translation):

> The views of space and time which I wish to lay before you have sprung from the soil of experimental physics, and therein lies their strength. They are radical. Henceforth space by itself, and time by itself, are doomed to fade away into mere shadows, and only a kind of union of the two will preserve an independent reality.

---

[3]No, the term "physicist" was not in use yet. Should you travel back in time by more than a century and develop some sort of medical ailment, be sure you know the true expertise of any physician you encounter.

[4]Minkowski was 44 years old at the time of his address, and would die of appendicitis in January 1909.

Minkowski's address then continued to lay the foundations for the work done earlier on in this chapter, though it should be noted that the usefulness of imaginary numbers had yet to be realized; Minkowski subtracted the square of time, not because he treated time as imaginary, but because it was already known that this was an invariant quantity. Still, he laid a geometric foundation that is incredibly useful today: the space-time continuum.

The difficulty in graphing problems in relativity is that time can be treated as an imaginary fourth direction. We can graph two dimensions easily, and can approximate graphing three dimensions by "projecting" them onto a two dimensional page. Graphing four dimensions cannot be done on two dimensional paper with pen or pencil.[5] Minkowski found a means to graph space and time together without losing the unique qualities of time. He realized that we could not, as is traditional, keep all of our axes at right angles. These so-called Minkowski diagrams formed a picture to go along with the concepts, and opened the gates to allow physicists to envision the situations in a useful way. They could graph situations as measured by two different observers on a single graph, making their work far more efficient. This also added another tool to the toolkit.

With space and time this closely connected, and with multiple observers graphed on the same diagram, an even more startling discovery was made: there is a rotation that allows one to rotate from a spatial direction into a time direction! This rotation manifests itself physically in an unusual fashion: accelerating to a new speed is mathematically identical to rotating with respect to the time access. The "rotation" exists, and we do have the ability to implement it, but the imaginary nature of the time direction means it "looks" completely different than a spatial rotation to us.[6] We still cannot travel backwards in time, but Minkowski's work laid foundations to explain that as well.

3.3.1   Causality

It is an axiom[7] of science that cause precedes effect. Relativity seems to support this concept.

Imagine two events in the space-time continuum. One happens in a given place and time, and the other happens in a different place at a different time. One can calculate the invariant distance between these two events, in the manner described above. These calculations involve both positive and negative quantities, and the results can be positive, negative or zero. We say that the events are *time-like separated* in one case, *space-like separated* in another, and *light-like separated* in the third. Time-like separated events are those which can be connected by cause and

---

[5]Approximations can be made by creating multiple three dimensional graphs, each at a different time, but this is cumbersome and makes the time evolution of the system more difficult to see from the graph alone.

[6]It has been proposed that there are life forms which perceive time no differently than they perceive space, though no concrete evidence exists. It has also been proposed that said life forms live inside an artificial wormhole and are worshipped by the inhabitants of a nearby world that look identical to humans aside from their noses, so the science involved is highly questionable.

[7]An axiom is an idea that is accepted as true, but which cannot be formally proven. They typically haven't been formally disproved, either, as that would result in the loss of the axiom.

effect, while space-like separated events cannot. Light-like separated events form the border between these two categories.

Let us examine explicit examples. Imagine one drops a rubber ball on the floor, and you consider the two events in question to be the first time it hits the floor, and the second time it hits the floor (after bouncing.) These events are time-like separated because they happen at the same point in space; it is only time that separates them. The cause-effect relationship is possible and established by observation. Any two time-like separated events find that the square of the distance between them in the time direction is greater than the sum of those in the spatial directions. As a result, information about one event can travel at the speed of light at reach the spatial coordinates of the second event *before* the second event takes place.

Now imagine two people are bouncing balls, one on planet Earth, and the other on the rock formerly considered a planet which is named Pluto. If an observer in a spacecraft half way between them believes they are bouncing their balls simultaneously, then this observer will say that there is a separation in space, but not in time. These events are space-like separated; neither person can have any effect on the other person's ball bouncing activity. There is a difference in space, but not in time. Information travelling at the speed of light can leave one event the instant it happens, but by the time it reaches the spatial coordinates of the second event, the event has already happened. There can be no cause-effect relationship.

In the case of light-like separated events, the cause-effect relationship is possible, but the calculation of the four dimensional distance between the two events works out to exactly zero.

These ideas were not terribly new. What was new is the mathematical implications that seemed obvious in one of Minkowski's graphs: events that are time-like separated from the perspective of one observer are time-like separated from the perspectives of all possible observers. Similarly, space-like separations are always space-like, and light-like separations are always light-like. The implication supports the idea that cause and effect relationships must be preserved in space and time, and that "time travel"[8] is impossible. This idea will be revisited when gravity comes into play in a later chapter.

Now, for the nitty-gritty mathematical details.

3.4   Coordinate Axes in Euclidean Space

Newton and his contemporaries worked in "Euclidean space," meaning that all geometry could be defined in the context of neat, perpendicular directions, as described in near totality by the ancient Greek mathematician named Euclid. Initially, we work only in Euclidean space. If your exposure to geometry is limited to what you learned in a public school system, then odds are exceptional that this is the only type of space and geometry you've ever been exposed to.

---

[8]Here the term "time travel" is used to mean travelling from an arbitrary point in time to another, equally arbitrary point in time, with little or no perceived time lapse between them.

The definitions of "directions" above amounts to different coordinate axes, each of which has a basis vector associated with it. The discussion of "complete cancellation" refers to the linear independence of the different vectors involved. The typical basis vectors are defined as follows:

$$\mathbf{e}_x = \begin{bmatrix} 1 \\ 0 \\ 0 \end{bmatrix}$$

$$\mathbf{e}_y = \begin{bmatrix} 0 \\ 1 \\ 0 \end{bmatrix}$$

$$\mathbf{e}_z = \begin{bmatrix} 0 \\ 0 \\ 1 \end{bmatrix}$$

These are certainly not the only definitions. In linear algebra terms, a set of vectors are basis vectors if they span the space, meaning that an arbitrary vector $\mathbf{v}$ can be expressed as a sum of the multiples of these basis vectors. (i.e. you can "build" any vector using only the basis vectors.) Thus, they don't actually need to be orthogonal to each other, provided they don't overlap completely.

For example,

$$\mathbf{e}_1 = \begin{bmatrix} 1 \\ 0 \\ 0 \end{bmatrix}$$

$$\mathbf{e}_2 = \begin{bmatrix} 0 \\ 1 \\ 0 \end{bmatrix}$$

$$\mathbf{e}_3 = \begin{bmatrix} 0 \\ 0 \\ 1 \end{bmatrix}$$

is a valid set of basis vectors but

$$\mathbf{e}_1 = \begin{bmatrix} 1 \\ 1 \\ 0 \end{bmatrix}$$

$$\mathbf{e}_2 = \begin{bmatrix} 1 \\ 1 \\ 1 \end{bmatrix}$$

$$\mathbf{e}_3 = \begin{bmatrix} 0 \\ 0 \\ 1 \end{bmatrix}$$

is not (since the middle vector is a combination of the other two.) The most convenient basis vectors to work with are those that are *orthonormal* to each other, meaning that

$$\mathbf{e}_i \cdot \mathbf{e}_j = \begin{cases} 1, i = j \\ 0, i \neq j \end{cases}$$

3.4.1 Rotations

The length of a vector $|\mathbf{v}|$ can be calculated by $|\mathbf{v}| = \sqrt{\mathbf{v} \cdot \mathbf{v}}$, also called the norm of a vector. (This is where the "norm" part came from in "orthonormal.") A matrix R represents a rotation if, and only if, it changes neither the lengths of vectors it acts on nor the angles between vectors it acts on. These two conditions can be written mathematically as follows:

1. $|\mathbf{v}| = |R\mathbf{v}|$ for every possible $\mathbf{v}$.

2. $\mathbf{u} \cdot \mathbf{v} = (R\mathbf{v})(R\mathbf{v})$ for every possible $\mathbf{u}$ and $\mathbf{v}$.

The second condition works because the dot product between two vectors depends solely upon the lengths of those vectors and the angle between them. If we know the lengths haven't changed and the dot product hasn't changed, then the angle between the two vectors couldn't possibly have changed. To further explore this second condition, we note that

$$\mathbf{u} \cdot \mathbf{v} = \mathbf{u}^T \mathbf{v}$$

where the superscript T indicates that we are taking the transpose of that vector. So, if

$$\mathbf{u} = \begin{bmatrix} u_1 \\ u_2 \\ u_3 \end{bmatrix}$$

then

$$\mathbf{u}^T = \begin{bmatrix} u_1 & u_2 & u_3 \end{bmatrix}$$

With this in mind, our second condition reduces to the following:

$$(R\mathbf{u}) \cdot (R\mathbf{v}) = (R\mathbf{u})^T (R\mathbf{v}) = \mathbf{u}^T R^T R\mathbf{v} = \mathbf{u}^T \mathbf{v}$$

or

$$R^T R = I \tag{3.1}$$

which means

$$R^T = R^{-1}$$

It can be shown (but won't be shown here) that this single condition is equivalent to both of the above conditions. If $R^T = R^{-1}$, then R represents a rotation in Euclidean space.

3.5 Coordinate Axes in Non-Euclidean Space

The revelation that time is best measured with imaginary numbers has led to another startling discovery: the geometry of the world we live in is not Euclidean. Sure, the spatial dimensions are, but once you introduce time, that goes out the window.

When dealing with a non-Euclidean space, one requires new definitions of vectors and the dot product. Rather than use the traditional vectors and matrices, we branch out into *tensors*. Tensors differ from the familiar objects in a very basic way: they are organized collections of variables, rather than organized collections of numbers. So, the component of a position tensor which may be treated as the time coordinate in one reference frame transforms into the time coordinate in all reference frames. The *value* of this coordinate may change drastically from one reference frame to another, but the *meaning* of that coordinate remains unchanged.

Before moving into explicit examples, notation will be confirmed: all objects in Euclidean space will be surrounded with square brackets [], while tensor objects in non-Euclidean space will be surrounded with round brackets (). Euclidean vectors will appear in bold face such as $\mathbf{v}$, while their four dimensional tensor counterparts will appear with a vector arrow above as $\vec{v}$. Components of Euclidean objects are marked with Latin indices such as u_i, while non-Euclidean tensors are marked with Greek indices such as u_α. Euclidean indices can take on the values 1, 2 or 3, while non-Euclidean tensors use 1, 2 and 3 for spatial coordinates and 0 for time.

3.5.1   Dot Products and the Metric

In non-Euclidean space, it is frequently impossible to use an orthonormal basis of vectors. Thus, the dot products taken in non-Euclidean space must be adapted in some way. In special relativity, there are a few options, and all depend upon the definition of the metric tensor. To begin with, instead of the "length" of a vector, we discuss the "length" of the interval. This is identical to the above definition of a Euclidean norm, save for the fact that it remains squared to avoid the introduction of imaginary numbers.

The metric tensor $g_{\mu\nu}$ is the mathematical object that describes the shape of the geometry being used, represented by interactions between different basis vectors.[9] Its components are formed by taking the dot products of the basis vectors of the geometry. There are four conventions to the metric and tensors in special relativity.

Convention One: Imaginary time, Euclidean metric

One convention (preferred by Stephen Hawking, for example) is to use the identity matrix as the metric, just as is done in Euclidean geometry. Doing so means that the four components of a tensor are different types of components; the time component is strictly imaginary, while other components are strictly real. Thus, for a position

---

[9]These, technically, exist in Euclidean geometry as well, but using metrics in Euclidean geometry amounts to multiplying by the identity matrix in most cases, which makes utterly no difference at the end of the day.

tensor $\vec{x}$, we have

$$\vec{x} \cdot \vec{x} = \begin{pmatrix} ict & x & y & z \end{pmatrix} \begin{pmatrix} 1 & 0 & 0 & 0 \\ 0 & 1 & 0 & 0 \\ 0 & 0 & 1 & 0 \\ 0 & 0 & 0 & 1 \end{pmatrix} \begin{pmatrix} ict \\ x \\ y \\ z \end{pmatrix}$$

$$= \begin{pmatrix} ict & x & y & z \end{pmatrix} \begin{pmatrix} ict \\ x \\ y \\ z \end{pmatrix}$$

$$= -c^2 t^2 + x^2 + y^2 + z^2$$

Why, do you ask, is the time component ict and not just it? This is because the units must be consistent for all components of a tensor (or vector). To transform the units of time into units of space, we multiply by the speed of light. This is related to the unpopularity of this convention; people expect consistency from one component to another. Mathematically, they need to have the same units and be in the same set of numbers. While this convention puts all components in the set of complex numbers, some people still find it aesthetically displeasing to find that manifesting itself as a purely imaginary component and three purely real components.

6.1.2 Convention Two: Real tensor components, negative time intervals

The second convention (preferred by Albert Einstein and this author) is to use strictly real values for all components of tensors, and encode the imaginary nature of time in the g_{00} component of the metric. This is done by defining the metric tensor as

$$g_{\mu\nu} = \begin{pmatrix} -1 & 0 & 0 & 0 \\ 0 & 1 & 0 & 0 \\ 0 & 0 & 1 & 0 \\ 0 & 0 & 0 & 1 \end{pmatrix}$$

such that the dot product of the position vector becomes

$$\vec{x} \cdot \vec{x} = \begin{pmatrix} ct & x & y & z \end{pmatrix} \begin{pmatrix} -1 & 0 & 0 & 0 \\ 0 & 1 & 0 & 0 \\ 0 & 0 & 1 & 0 \\ 0 & 0 & 0 & 1 \end{pmatrix} \begin{pmatrix} ct \\ x \\ y \\ z \end{pmatrix}$$

$$= \begin{pmatrix} -ct & x & y & z \end{pmatrix} \begin{pmatrix} ct \\ x \\ y \\ z \end{pmatrix}$$

$$= -c^2 t^2 + x^2 + y^2 + z^2$$

In this case, events that satisfy the rules of causality (i.e. information from event 1 can reach event 2 by travelling at or below the speed of light) have negative lengths, or norms. This is identical to the convention using imaginary time, but allows all tensor components to be real numbers at the cost of having a nontrivial metric.

3.5.2   Convention Three: Real tensor components, positive time intervals

The third convention, which may be the most popular convention, is to define the metric so that causal events are separated by positive intervals. Doing this requires a sign change in the entire metric:

$$
g_{\mu\nu} = \begin{pmatrix} 1 & 0 & 0 & 0 \\ 0 & -1 & 0 & 0 \\ 0 & 0 & -1 & 0 \\ 0 & 0 & 0 & -1 \end{pmatrix}
$$

resulting in a sign change all the way through the dot product:

$$
\vec{x} \cdot \vec{x} = \begin{pmatrix} ct & x & y & z \end{pmatrix} \begin{pmatrix} 1 & 0 & 0 & 0 \\ 0 & -1 & 0 & 0 \\ 0 & 0 & -1 & 0 \\ 0 & 0 & 0 & -1 \end{pmatrix} \begin{pmatrix} ct \\ x \\ y \\ z \end{pmatrix}
$$

$$
= \begin{pmatrix} ct & -x & -y & -z \end{pmatrix} \begin{pmatrix} ct \\ x \\ y \\ z \end{pmatrix}
$$

$$
= c^2 t^2 - x^2 - y^2 - z^2
$$

This means that most tensors of interest have positive intervals, but the spatial components take on negative values when squared. This is a perfectly valid convention, but one that is often found to be counterintuitive in this manner. Still, most people seem to have an easier time with negative space components than negative intervals, and so it is very popular.

3.5.3   Convention Four: Quaternions

This fourth and least popular convention is a blend of conventions one and three. In this one, tensors are represented with quaternions, which are essentially complex numbers with three imaginary components. Thus, the metric remains Euclidean, and the mix of real and imaginary numbers feels more natural as it comes in an established mathematical form. The dot product looks like this:

$$
\vec{x} \cdot \vec{x} = \begin{pmatrix} ct & ix & jy & kz \end{pmatrix} \begin{pmatrix} 1 & 0 & 0 & 0 \\ 0 & 1 & 0 & 0 \\ 0 & 0 & 1 & 0 \\ 0 & 0 & 0 & 1 \end{pmatrix} \begin{pmatrix} ct \\ ix \\ jy \\ kz \end{pmatrix}
$$

$$
= \begin{pmatrix} ct & ix & jy & kz \end{pmatrix} \begin{pmatrix} ct \\ ix \\ jy \\ kz \end{pmatrix}
$$

$$
= (ct)^2 + (ix)^2 + (jy)^2 + (kz)^2
$$

$$
= c^2 t^2 - x^2 - y^2 - z^2
$$

This text will use convention two, using entirely real components to a tensor, but with a non-Euclidean metric, due in part to the fact that non-Euclidean metrics will be inescapable in chapters 7 through 9.

3.6 Basic Operations with Matrices and Tensors

3.6.1 Matrices

Matrices can be added, subtracted, multiplied, and (after a fashion) divided, subject to certain rules. They are rectangular in shape, with numbers arranged in rows and columns. When multiplying matrices, order matters. In other words, if the two matrices are A and B, then $AB \neq BA$ in most cases. In fact, it may not even be possible to calculate AB when BA is perfectly well defined. For example, let

$$A = \begin{bmatrix} 1 & 1 & 1 \\ 2 & 2 & 2 \\ 3 & 3 & 3 \end{bmatrix}$$

and

$$B = \begin{bmatrix} 4 & 7 \\ 5 & 8 \\ 6 & 9 \end{bmatrix}$$

To calculate the number in row i and column j of AB, we multiply each entry in row i of matrix A by each entry in column j of matrix B and add them up. Thus,

$$AB = \begin{bmatrix} 1\cdot4+1\cdot5+1\cdot6 & 1\cdot7+1\cdot8+1\cdot9 \\ 2\cdot4+2\cdot5+2\cdot6 & 2\cdot7+2\cdot8+2\cdot9 \\ 3\cdot4+3\cdot5+3\cdot6 & 3\cdot7+3\cdot8+3\cdot9 \end{bmatrix}$$
$$= \begin{bmatrix} 15 & 24 \\ 30 & 48 \\ 45 & 72 \end{bmatrix}$$

We can do this because the number of columns in A is the same as the number of rows in B. We cannot calculate BA: there are only two entries in the first row of B, but there are three entries in the first column of A. A has no entries to pair with the bottom row of B. When multiplication of two arbitrary matrices A and B is possible, the resultant matrix has the same number of rows as A and the same number of columns as B. In general, if A has n rows and m columns ("A is $n \times m$"), and B has r rows and s columns ("B is $r \times s$"), then AB is defined if and only if $m = r$, and matrix AB will have n rows and s columns ("AB is $n \times s$"). The entry in row i and column j of matrix A is denoted A_{ij}.

Matrices of all sizes have an algebraic equivalent of zero: every entry in the matrix is a zero. Not all sizes of matrix have an algebraic equivalent of one. The only matrices that behave algebraically like the number one are the identity matrices we call I. I behaves as $AI = IA = A$ no matter what A is. This must be the same I when multiplied by either the left or the right, so (by the rules of multiplication) it must be a square ($n \times n$ for some n) matrix. If I is the identity matrix, then $I_{ij} = 1$

if $i = j$ and $I_{ij} = 0$ if $i \neq j$. Visually, this is a matrix that has the number 1 appear in every entry of the diagonal from the top left corner to the bottom right corner, with the number 0 everywhere else. Matrices cannot be easily divided. Square matrices can approximate this in certain cases. Some square matrices (but not all) have an inverse; matrix C is the inverse of matrix B if $CB = BC = I$. C is usually denoted B^{-1} if it exists, and it will have the same size as B. Details about which matrices have inverses and which do not will not be provided here; search online for "determinant" to find the answer to that question.

Matrices can be added and subtracted, as well, provided the two matrices involved are of exactly the same size, whatever size that is. If A is $n \times m$, then A can only be added to B if B is also $n \times m$. The basic operations with matrices, if defined, can be generally expressed as follows:

$$(A + B)_{ij} = A_{ij} + B_{ij}$$
$$(A - B)_{ij} = A_{ij} - B_{ij}$$
$$(AB)_{ij} = \sum_k A_{ik} \cdot B_{kj}$$
$$(A \div B)_{ij} = AB^{-1} = \sum_k A_{ik} \cdot (B^{-1})_{kj}$$

The dot product of two vectors $\mathbf{u}$ and $\mathbf{v}$ can then be written as

$$\sum_i \sum_j g_{ij} u_i v_j = \sum_i u_i v_i$$

where we have used the Euclidean metric

$$g_{ij} = \begin{cases} 1, i = j \\ 0, i \neq j \end{cases}$$

3.6.2   Tensors

There are comparable situations for tensors, though the indices vary somewhat. When a tensor fills the role of a four dimensional vector, it is called a *four-vector*, and its index is raised, as with $\bar{x}^\mu$ for the position four-vector. When written out in matrix form within curved brackets, they perform in the same manner as Euclidean vectors. The differences lie in the fact that indices can be raised or lowered due to the non-Euclidean metric. The dot product of two four-vectors $\vec{u}$ and $\vec{v}$ is given by

$$\sum_\mu \sum_\nu g_{\mu\nu} u^\mu v^\nu = \sum_\nu u_\nu v^\nu$$

Notice that the indices move when summed over an index with the metric tensor. This process transforms a four-vector into a *one-form*, which has the lowered index. In its explicit form, this amounts to taking the transpose of $\vec{u}$ and multiplying it through the metric from the left as follows:

$$u_\nu = \sum_\mu g_{\mu\nu} u^\mu = \begin{pmatrix} u_0 & u_1 & u_2 & u_3 \end{pmatrix} \begin{pmatrix} -1 & 0 & 0 & 0 \\ 0 & 1 & 0 & 0 \\ 0 & 0 & 1 & 0 \\ 0 & 0 & 0 & 1 \end{pmatrix}$$

$$u_\nu = \begin{pmatrix} -u_0 & u_1 & u_2 & u_3 \end{pmatrix}$$

When the indices are not explicitly written, there must be some way to distinguish between four-vectors and one-forms. Just as $u^\mu = \vec{u}$ is the four-vector notation, $u_\mu = \tilde{u}$ is the one-form notation.

There will be a lot of summations over indices in the upcoming chapters, so we adopt the *Einstein summation convention*: if the same letter appears as an index that is both raised and lowered in a term, then the sum over that index is implied. This reduces the writing in our dot product from

$$\sum_\mu \sum_\nu g_{\mu\nu} u^\mu v^\nu = \sum_\nu u_\nu v^\nu$$

to

$$g_{\mu\nu} u^\mu v^\nu = u_\nu v^\nu$$

which saves a lot of writing and typing. Similarly, we can write $\vec{u} \cdot \vec{v} = \tilde{u}\vec{v}$, in which the order of the terms matters; $\vec{v}\tilde{u}$ is a very different quantity. (In matrix form, the former becomes a single scalar quantity, while the latter becomes a $4 times 4$ matrix of its own.) In the case of the metric of special relativity itself, as long as both indices are together, we have the above metric: $g_{\mu\nu} = g^{\mu\nu}$. In the case of mixed indices, such as g^μ_ν, then we have the identity matrix.

Tensors are most interesting when one transforms them from one reference frame to another.

3.7   The Lorentz Boost Tensor

The Lorentz boost tensor $\Lambda^{\mu'}{}_\nu$ which transforms the position four-vector

$$x^\nu = \begin{pmatrix} ct \\ x \\ y \\ z \end{pmatrix}$$

as measured in reference frame S into the position four-vector

$$x^{\mu'} = \begin{pmatrix} ct' \\ x' \\ y' \\ z' \end{pmatrix} = \begin{pmatrix} \gamma \left(ct - \frac{v}{c}x \right) \\ \gamma \left(x - vt \right) \\ y \\ z \end{pmatrix}$$

as measured in reference frame S' is given by

$$\Lambda^{\mu'}{}_{\nu} = \begin{pmatrix} \gamma & -\frac{v}{c}\gamma & 0 & 0 \\ -\frac{v}{c}\gamma & \gamma & 0 & 0 \\ 0 & 0 & 1 & 0 \\ 0 & 0 & 0 & 1 \end{pmatrix}$$

where we continue to use the standard

$$\gamma = \frac{1}{\sqrt{1 - \frac{v^2}{c^2}}}$$

Thankfully, it can (but won't) be shown that this is the same tensor that transforms all tensors from frame S to frame S', where the $'$ symbol is placed on the indices to indicate which are and are not in frame S'.

3.7.1   Lorentz Boost as a Rotation

In Euclidean space, a matrix R represents a rotation if it satisfies equation (3.1):

$$R^T R = I$$

In non-Euclidean space, the condition is slightly different:

$$R^T g R = g$$

In other words, the metric must be preserved. We can show that this is the case for the Lorentz boost by multiplying out the matrix forms:

$$\Lambda^T g \Lambda = \begin{pmatrix} \gamma & -\frac{v}{c}\gamma & 0 & 0 \\ -\frac{v}{c}\gamma & \gamma & 0 & 0 \\ 0 & 0 & 1 & 0 \\ 0 & 0 & 0 & 1 \end{pmatrix} \begin{pmatrix} -1 & 0 & 0 & 0 \\ 0 & 1 & 0 & 0 \\ 0 & 0 & 1 & 0 \\ 0 & 0 & 0 & 1 \end{pmatrix} \begin{pmatrix} \gamma & -\frac{v}{c}\gamma & 0 & 0 \\ -\frac{v}{c}\gamma & \gamma & 0 & 0 \\ 0 & 0 & 1 & 0 \\ 0 & 0 & 0 & 1 \end{pmatrix}$$

$$= \begin{pmatrix} -\gamma & -\frac{v}{c}\gamma & 0 & 0 \\ \frac{v}{c}\gamma & \gamma & 0 & 0 \\ 0 & 0 & 1 & 0 \\ 0 & 0 & 0 & 1 \end{pmatrix} \begin{pmatrix} \gamma & -\frac{v}{c}\gamma & 0 & 0 \\ -\frac{v}{c}\gamma & \gamma & 0 & 0 \\ 0 & 0 & 1 & 0 \\ 0 & 0 & 0 & 1 \end{pmatrix}$$

$$= \begin{pmatrix} -\gamma^2\left(1 - \frac{v^2}{c^2}\right) & 0 & 0 & 0 \\ 0 & \gamma^2\left(1 - \frac{v^2}{c^2}\right) & 0 & 0 \\ 0 & 0 & 1 & 0 \\ 0 & 0 & 0 & 1 \end{pmatrix}$$

$$= \begin{pmatrix} -1 & 0 & 0 & 0 \\ 0 & 1 & 0 & 0 \\ 0 & 0 & 1 & 0 \\ 0 & 0 & 0 & 1 \end{pmatrix}$$

$$= g$$

3.7.2 Invariants

The invariant quantities between reference frames are those whose magnitude does not change. In other words, a quantity is invariant if and only if

$$\vec{u} \cdot \vec{w} = \vec{u}' \cdot \vec{w}'$$

or

$$u_\nu w^\nu = g_{\mu\nu} u^\mu w^\nu = g_{\alpha'\beta'} u^{\alpha'} w^{\beta'} = u_{\beta'} w^{\beta'}$$

Well, the right hand side can be rewritten as follows:

$$= \Lambda^\nu{}_{\beta'} g_{\mu\nu} u^\mu w^{\beta'}$$
$$= g_{\mu\nu} u^\mu w^\nu$$
$$= u_\nu w^\nu$$

which now matches the left hand side. Thus, all tensors are automatically invariant. In fact, the formal definition of a tensor which distinguishes it from Euclidean objects is that it *does* maintain an invariant interval after a Lorentz transformation.

To see this is all its tedious, matrix form glory, we have:

$$\vec{u}' \cdot \vec{w}' = \begin{pmatrix} u_0 & u_1 & u_2 & u_3 \end{pmatrix} \begin{pmatrix} \gamma & -\frac{v}{c}\gamma & 0 & 0 \\ -\frac{v}{c}\gamma & \gamma & 0 & 0 \\ 0 & 0 & 1 & 0 \\ 0 & 0 & 0 & 1 \end{pmatrix} \times$$

$$\begin{pmatrix} -1 & 0 & 0 & 0 \\ 0 & 1 & 0 & 0 \\ 0 & 0 & 1 & 0 \\ 0 & 0 & 0 & 1 \end{pmatrix} \begin{pmatrix} \gamma & -\frac{v}{c}\gamma & 0 & 0 \\ -\frac{v}{c}\gamma & \gamma & 0 & 0 \\ 0 & 0 & 1 & 0 \\ 0 & 0 & 0 & 1 \end{pmatrix} \begin{pmatrix} w_0 \\ w_1 \\ w_2 \\ w_3 \end{pmatrix}$$

$$= \begin{pmatrix} u_0 & u_1 & u_2 & u_3 \end{pmatrix} \begin{pmatrix} -\gamma & -\frac{v}{c}\gamma & 0 & 0 \\ \frac{v}{c}\gamma & \gamma & 0 & 0 \\ 0 & 0 & 1 & 0 \\ 0 & 0 & 0 & 1 \end{pmatrix} \times$$

$$\begin{pmatrix} \gamma & -\frac{v}{c}\gamma & 0 & 0 \\ -\frac{v}{c}\gamma & \gamma & 0 & 0 \\ 0 & 0 & 1 & 0 \\ 0 & 0 & 0 & 1 \end{pmatrix} \begin{pmatrix} w_0 \\ w_1 \\ w_2 \\ w_3 \end{pmatrix}$$

$$= \begin{pmatrix} u_0 & u_1 & u_2 & u_3 \end{pmatrix} \begin{pmatrix} -\gamma^2(1 - \frac{v^2}{c^2}) & 0 & 0 & 0 \\ 0 & \gamma^2(1 - \frac{v^2}{c^2}) & 0 & 0 \\ 0 & 0 & 1 & 0 \\ 0 & 0 & 0 & 1 \end{pmatrix} \begin{pmatrix} w_0 \\ w_1 \\ w_2 \\ w_3 \end{pmatrix}$$

$$= \begin{pmatrix} u_0 & u_1 & u_2 & u_3 \end{pmatrix} \begin{pmatrix} -1 & 0 & 0 & 0 \\ 0 & 1 & 0 & 0 \\ 0 & 0 & 1 & 0 \\ 0 & 0 & 0 & 1 \end{pmatrix} \begin{pmatrix} w_0 \\ w_1 \\ w_2 \\ w_3 \end{pmatrix}$$

$$= \begin{pmatrix} -u_0 & u_1 & u_2 & u_3 \end{pmatrix} \begin{pmatrix} w_0 \\ w_1 \\ w_2 \\ w_3 \end{pmatrix}$$

$$= \vec{u} \cdot \vec{w}$$

Notice that, aside from the u and w terms, this looks very much like the proof that $\Lambda^\nu{}_{\beta'}$ is a rotation. This is a direct result of the invariance of a tensor under a Lorentz transformation.

3.7.3 Minkowski

Minkowski was the first to discover an effective way to create coordinate axes and graph multiple reference frames on a single graph. This was such a useful discovery that they are now known as Minkowski diagrams.

We start with the axes for reference frame S:

Note that time is on the vertical axis in a Minkowski diagram. Now, we can also insert a line that shows how a photon travels through the geometry of the universe:

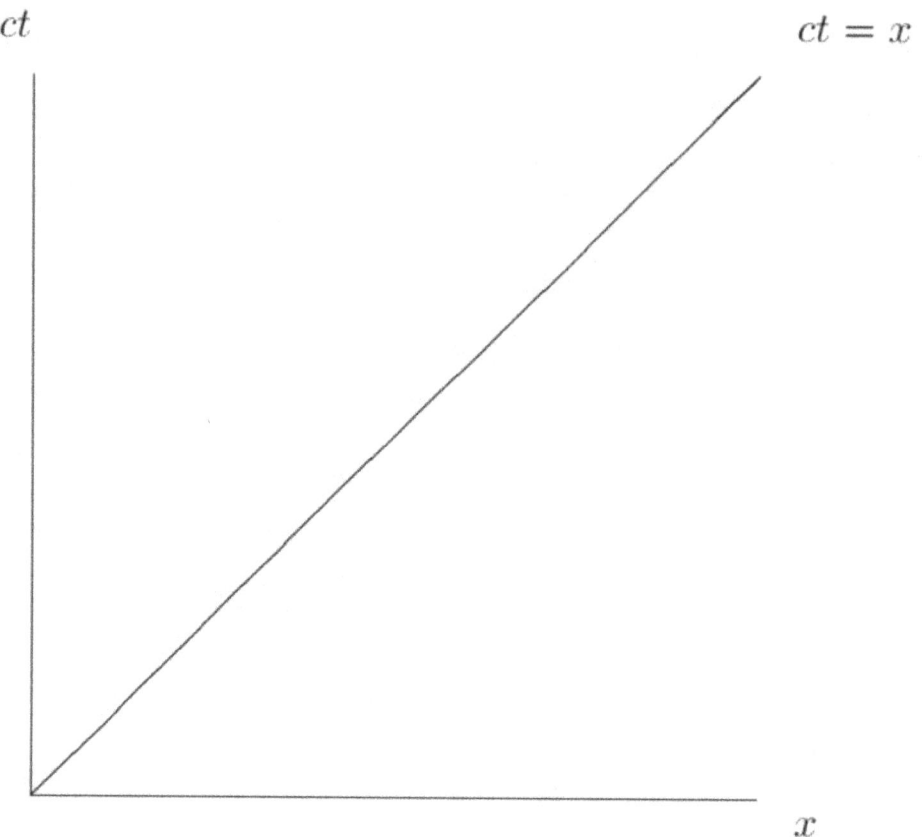

The new question becomes: how do we plot axes for events that take place in the reference frame S'? One axis is quite straightforward: if we watch the spatial origin of this reference frame, we find that it will move in the x direction with speed $v = \frac{x}{t}$, so that it makes a line where $x' = y' = z' = 0$, which forms the ct' axis. In frame S, this line meets the ct axis at an angle θ. Using the definition of the tangent function from trigonometry, we see that $\tan\theta = \frac{x}{ct} = \frac{v}{c}$; thus, the relative speed v of the two reference frames provides all the information required to plot the ct' axis:

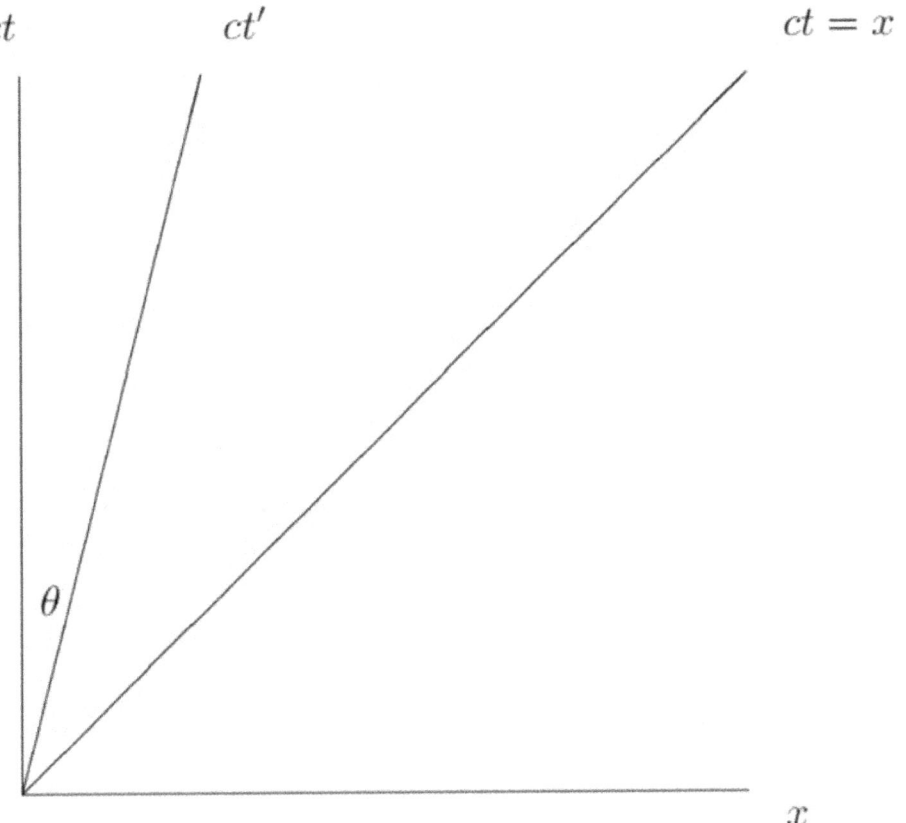

The question remains: where does the x' axis go? Well, one of the things every observers agrees on is orthogonality. If $\vec{u} \cdot \vec{w} = 0$, then $\vec{u}' \cdot \vec{w}'$. So, which line is orthogonal with the axis we have already drawn? It is the line at which $ct' = 0$. In other words, it is the locus of points that those in frame S perceive as forming the x' axis at the time $t' = 0$. These points can be plotted using the same logic as above, knowing how points transform over time. Somewhat surprisingly, this orthogonal axis doesn't *look* orthogonal the first time it is viewed:

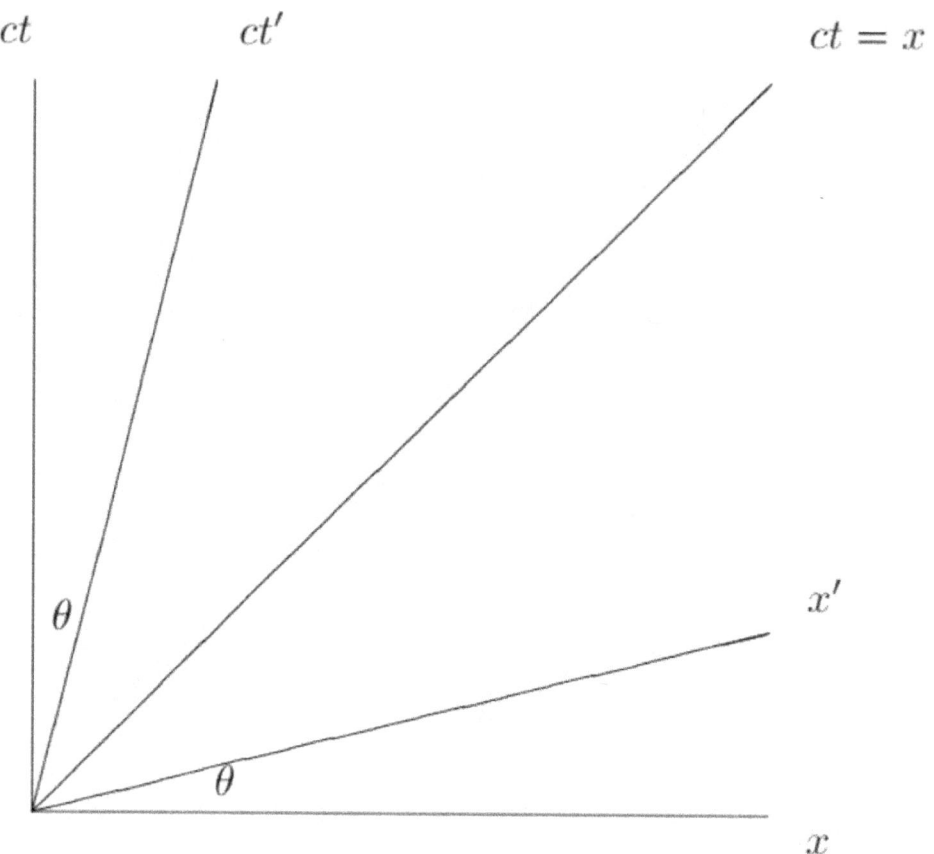

You don't need a protractor to see that the ct' and x' axes are not at a 90° angle to each other. Our exclusively Euclidean experience drives us to assume that right angles are the only form of orthogonality, but when time is measured on an imaginary axis, this is not the case. What matters is the angle made with respect to the $x = ct$ line, which also coincides with the $x' = ct'$ line. In fact, that link is the only line that remains invariant for every axis. It should be noted that this diagram assumes $v > 0$ m/s. If the value is negative, then the S' frame's axes are outside the S frame's axes, but the magnitude of the angle θ remains unchanged.

Chapter 4

Paradoxes of Relativity

4.1 Introducing the Paradoxes

When Einstein's theory of relativity was first published, its counter-intuitive implications were met with extreme levels of doubt and scepticism. Early researchers actively sought out logical contradictions in the framework that could be used to justify a wholesale rejection of the theory so that it might be replaced with something more aesthetically pleasing. Some of these "paradoxes" are described here. Study of the paradoxes is remarkably illuminating, as it forces the student to directly confront intuitions and prejudices developed from years of working with Newtonian mechanics.

4.2 Length Contraction

The length contraction paradox is one of the earliest paradoxes. Imagine you have a sports car that is 5 meters long, and your garage is only 4 meters deep. Well, if "moving objects contract," as Einstein's theory predicts, then you should be able to fit your sports car in your garage, provided you are moving quickly enough that the moving vehicle contracts to a length that is less than 4 meters. Everything is fine until you stop the vehicle.

The paradox comes from the equivalence of all reference frames. From the perspective of the driver, the car is stationary, and it is the garage that is in motion. In that case, the 5 meter car is being approached headlong by a garage that is less than 4 meters long: the garage is still smaller than the car. An observer standing near (or in) the garage, at rest with respect to the garage, concludes that the car fits in the garage; an observer in the car concludes that the car does not fit inside the garage. Now, the car is either in the garage or it is not, in all reference frames. So, which is it? This was one of the "thought experiments"[1] that seemed to cause

---

[1]A "thought experiment" is an experiment one thinks about but doesn't actually do. This causes a dramatic improvement in lab budgets, allows a tremendous opportunity to explore implications of a theory, and severely limits the ability to make quantified observations to verify a theory.

problems for Einstein's theory.

4.2.1   Relativity of Simultaneity

The key to resolving this paradox lies in something known as the *relativity of simultaneity*. Two events are considered simultaneous if they happen "at the same time." It is not immediately obvious to most people that this concept depends greatly upon how we measure time. If time is measured differently by two observers, will they agree on whether or not events are simultaneous?

Let us examine what we mean when we say two events happen "at the same time." This means that information about these two events reach the observer at the same time. To pick a more mundane example, picture two different patrons at an outdoor concert in a very large venue. The stage has a large screen at the rear with live video of the vocalist, and most of the speakers projecting sound to the audience are at the front. The patron near the stage believes that the sound and image are simultaneous, as the lips and lyrics are synchronized on screen. The patron at the rear of the venue believes that the sound is out of synchronization with the video because the light from the screen brings that patron information about the video a short time before the sound from the speakers reaches the patron. This is true in Newtonian mechanics as well, simply because light is faster than sound, so that information arrives faster. They do not appear to be simultaneous.[2]

In relativity, we can pretend all information travels at the speed of light. A somewhat analogous situation arises in this case. Events are simultaneous for an observer if the information about those two events arrives at the same time. Since all information travels at the speed of light, this amounts to saying they travelled the same distance to the observer. This is easy enough to imagine when nothing is in motion: the observer is exactly centred between them.[3] What if the observer is in motion? What if one or more events are in motion?

Imagine that the events take place at locations A and B. To an observer who is at rest relative to those locations and is exactly half way between them, the events are simultaneous. Now imagine an observer in motion from location A to location B. This moving observer will only consider the events to be simultaneous if she or he is the same distance (equidistant) from both locations when information about the two events arrives. So, if the moving observer is exactly on top of our stationary observer when the information arrives, both observers will agree that the events are simultaneous. However, if the moving observer is not in this symmetric position,

---

Thought experiments are best used to design experiments to be done physically later.

[2]This analogy, of course, derives from limitations of human perception. If we could accurately perceive time in smaller intervals, the patron closer to the stage would also recognize that the events are not simultaneous. The only observer who could say the two events are simultaneous would be the observer who eyes and ears are pressed up against the face of the vocalist. This is not a viewing location recommended to those who want to watch the entire concert without being dragged off by members of the venue's security force.

[3]Technically, the observer could be anywhere on the infinitely large plane between the two events. Imagine that the two events that occur are centred along the outside edges of the pages of an open book. The observer could be anywhere alone the spine where the pages meet. This book can be as tall as you'd like, and can be spun around. The observer can be any place the spine can be. This forms the infinite plane.

then information about the two events will arrive at two different times, and the moving observer will not consider the events to be simultaneous.

4.2.2 Resolving the Length Contraction Paradox

Somewhat contrary to intuition, the definition of simultaneity plays a large role in the length contraction paradox. Upon closer inspection, the paradox arises because the mental images it conjures for the driver involves knowledge about both ends of the car simultaneously. In other words, the transmission of information is considered instantaneous instead of delayed, violating the postulates of relativity. The solution to the paradox comes through in a manner that surprised the opponents of relativity: when the car enters the garage, the front of the car strikes the rear garage wall. Information about this collision is then transmitted[4] to the rear of the car. In the time it takes this information to reach the rear of the car, the rear of the car has already entered the garage;[5] both observers agree that the car fits in the garage! Both observers also agree that the driver is insane and the car is a write off.

4.3 Twin Paradox

The equivalence of reference frames also leads to problems with time dilation.

Imagine you are a twin.[6] At present, you are both the same age.

The paradox begins when you load your twin into a rocket ship and launch him or her on a round trip to another solar system a few light years away. Your twin is in motion the entire trip, so as far as you are concerned, time moves slower in her or his frame of reference. When the craft returns, you will be the older twin. Now imagine your twin's perspective. Your twin can consider himself or herself to be at rest the entire trip, so you should be aging more slowly, and your twin will be the older twin upon her or his return. Which of you is older?

4.3.1 Resolving the Twin Paradox

The key to this paradox is the recognition that we are not, in fact, dealing with inertial reference frames. Recall the definition from our previous chapter: an "inertial" reference frame is one in which the origin of the coordinate system is not being accelerated. Acceleration includes direction as well as speed. When the twin in the rocket turns around for the return trip, she or he changes from one inertial

---

[4]The mechanics of transmission are irrelevant to the paradox. In the physical world, the information is transmitted via vibrations in the bonds between atoms and molecules in the car, and is transmitted at less than the speed of light. A lower transmission speed assists in the resolution of the paradox, and will be ignored in the main argument.

[5]This is guaranteed. If you look carefully at the math, the condition that says the car is fast enough for length contraction to make the car fit in the garage from the perspective of the observer in the garage is identical to the mathematical condition that the rear of the car enter the garage before information about the collision arrives. The car fits or doesn't fit when measured by all observers.

[6]If you really are a twin, imagine this is one of the days that you and your twin are getting on each other's nerves.

reference frame to another. In reversing direction, the pretence of simultaneity is broken, and the world proceeds as usual.

Complete details are available in the version of this chapter that includes the math. The crux of the solution is this: when the twin is rocketing away from Earth[7] both are correct in treating themselves as the "older" twin. If each twin sends signals to the other at regular intervals (such as birthdays) then both twins send more signals than they receive while they are moving apart. When the twin in the rocket reverses course, however, the timing is disrupted. This twin begins to see signals more quickly as he or she approaches their point of origin. By the time both twins are back on their home world, the twin in the rocket has received more signals than she or he has sent, and they agree that the Earthbound twin is older.[8] This does leave one more question, though: what is the age of the twin who made the round trip? For that, we measure the proper time. Proper time is not "more correct" than any other time, despite the instinctive reaction to the name. It is the time as measure in the rest frame of the moving twin. This is the amount of time that lines up properly with that twin's natural aging process.

These are not the only "paradoxes" of relativity that one can find. Thankfully, all are resolvable: every seeming paradox proposed to date has been found to be reconcilable, and most often "arise" by applying our normal, day-to-day intuitions which are actually in error, rather than the theory of relativity itself.

4.4 Length Contraction Paradox: Mathematical Solution

There are two possible approaches to solving this problem. One is to do three pages of algebra, calculating the observed lengths of both objects in both reference frames and then comparing with the time lag. The other way is to use a single Minkowski diagram. We shall use the latter approach.

We begin by drawing our axes, taking the garage as frame S and the car as being at rest in frame S':

---

[7]Assuming our readers are on Earth. If you are reading this on another planet, please e-mail me at `w.blaine.dowler@gmail.com` and let me know!

[8]This is punishment for sending your twin on a long term interstellar trip, I suppose.

ct

x

Let the garage be two units wide:

Now we let the car be three units wide in frame S', and travelling at $v = \frac{5}{6}c$. This contracts it to 2 units long in frame S'. We plot the front of the car up to and including the moment it hits the back wall of the garage:

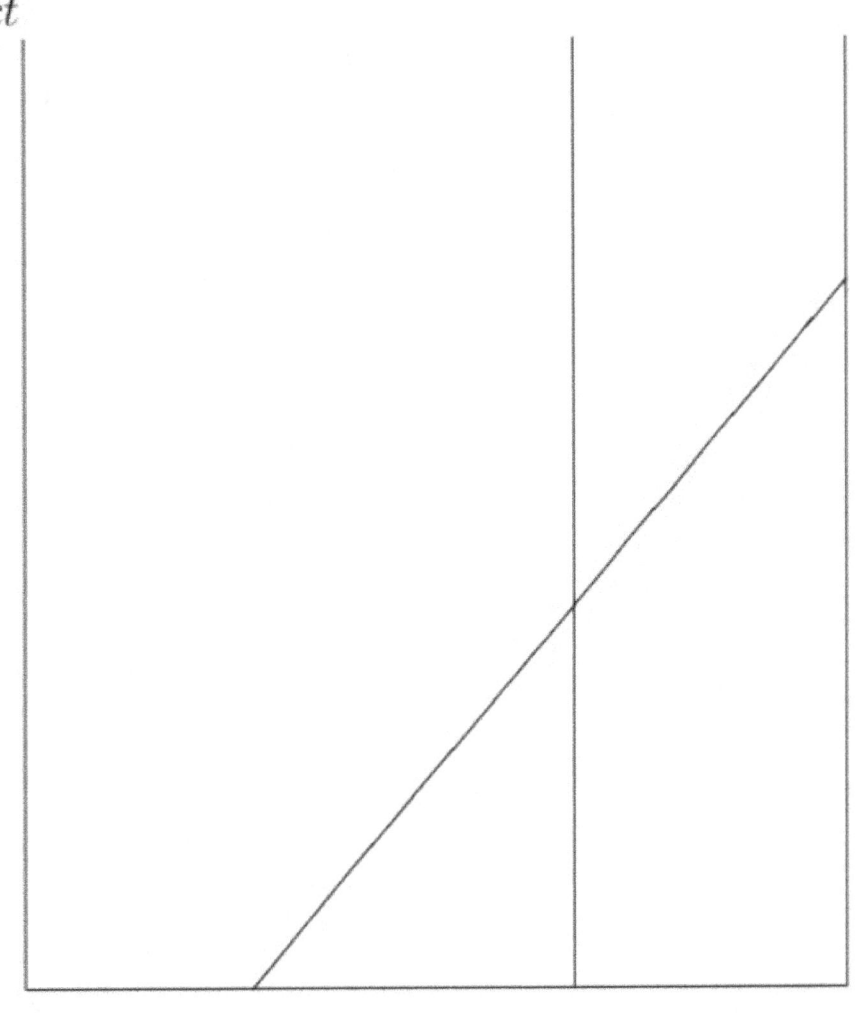

We can plot the rear of the car the same way. Simultaneously, we plot the "information" about the collision of the front of the car with the back of the garage, and run both lines until they intersect.

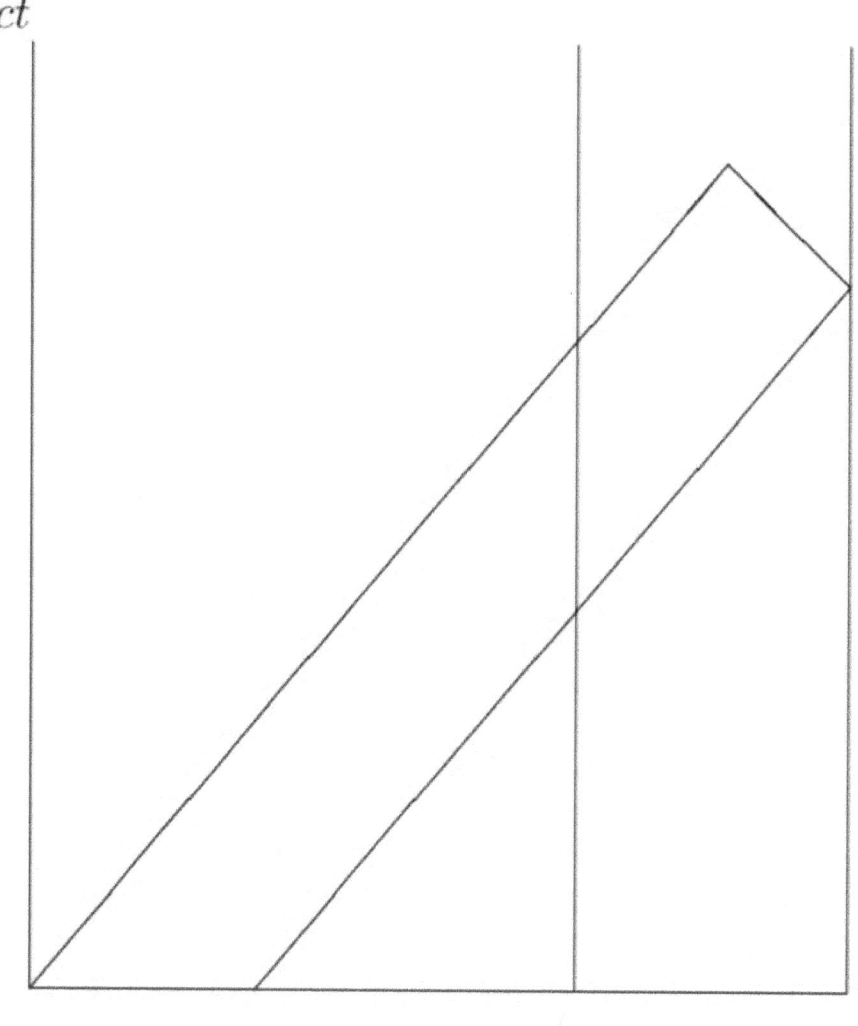

From the perspective of reference frame S, any points that occur along lines parallel to the x axis are simultaneous (same ct value), so a short horizontal line can be used to show that the entire car fits in the garage from that reference frame:

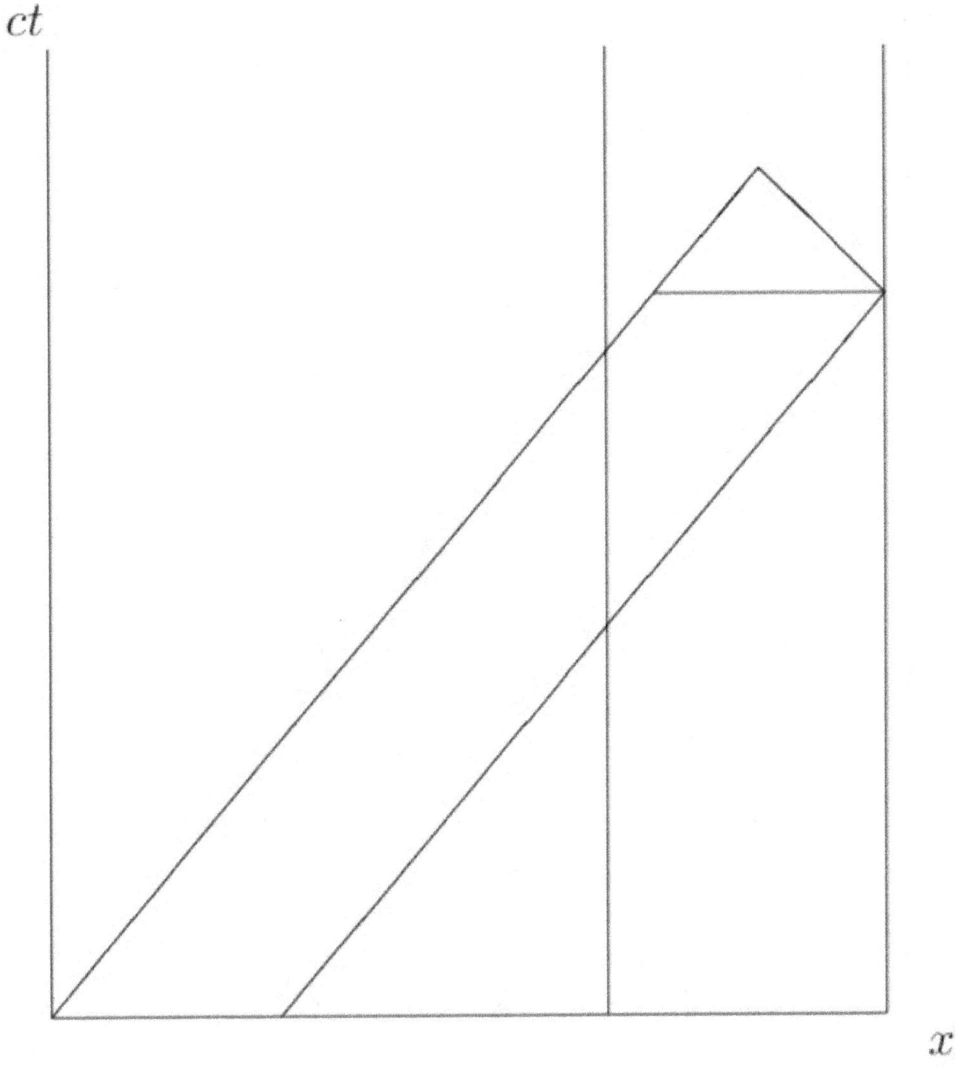

At the point of intersection, the back of the car is inside the garage from the perspective of reference frame S. From the perspective of reference frame S', the back
of the car is always sitting on the ct' axis, so we simply add the x' axis:

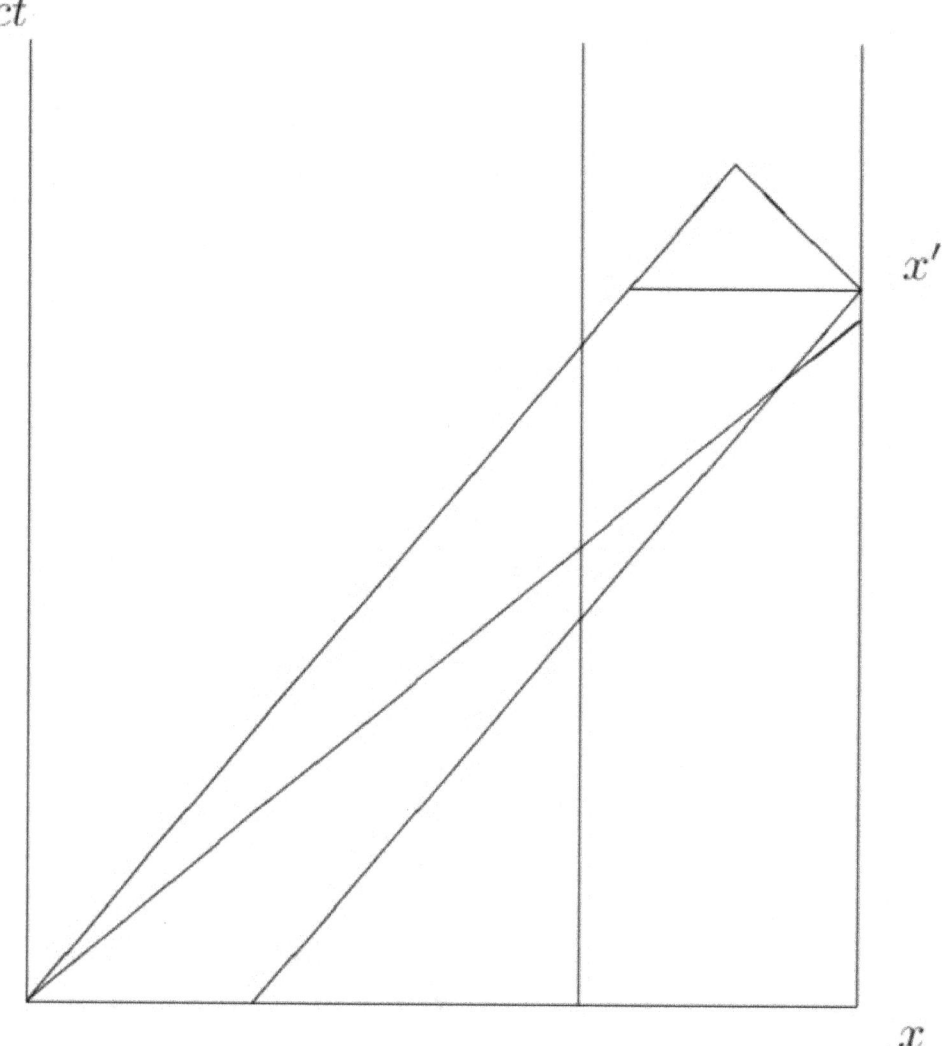

Now, to check frame S', we look at where the rear of the car "perceives" the front of the car to be at the point of collision. We do this by creating a line parallel to x' coming from the point at which the rear of the car and the front of the garage are in identical locations in space.

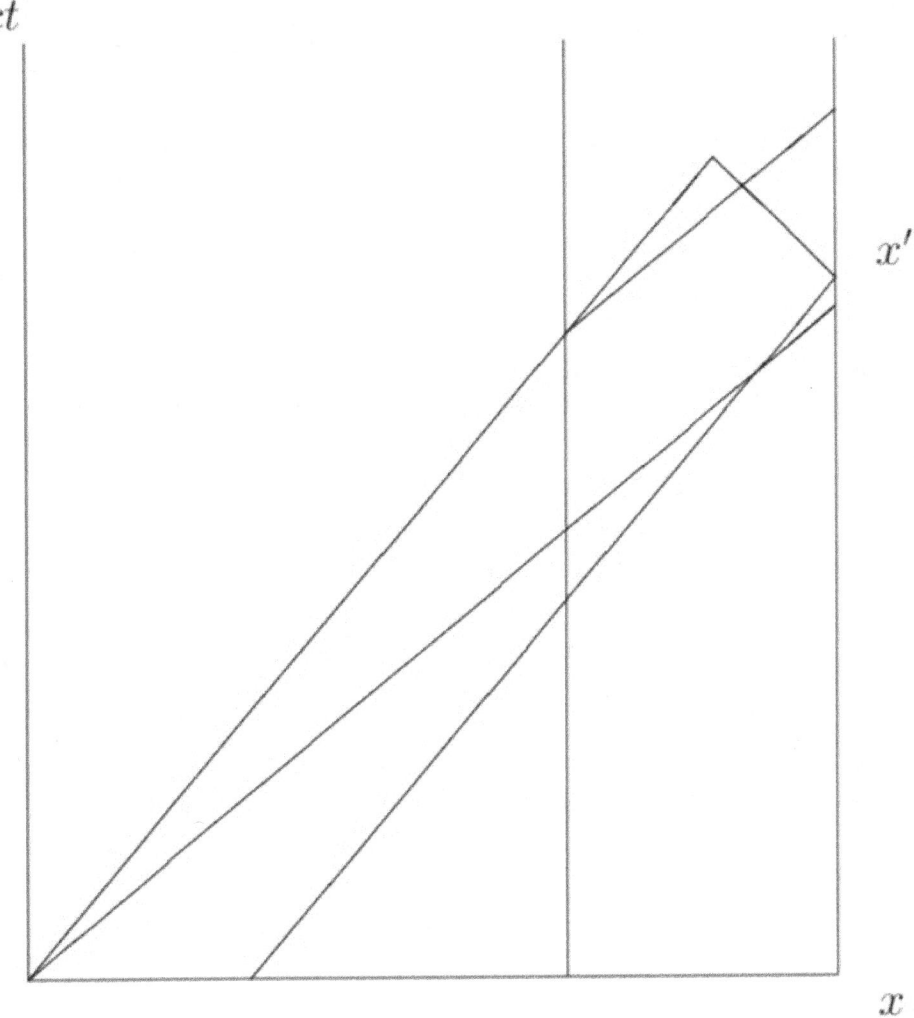

Note that this line is "below" the point of intersection between the rear of the car and the information about the collision at the front of the garage. Thus, due to the relativistic version of simultaneity, the rear of the car enters the garage before the collision in both reference frames.

4.5 Twin Paradox: Mathematical Solution

The key to resolving the twin paradox is to calculate the time elapsed in each half of the trip separately. This will automatically cover the change in reference frames.

There are three events in time of relevance. The first is the event of launch, when one twin remains on Earth and the other does not. At this point, we can set $ct = ct' = x = x' = 0 lyr$ by choice. (We will be measuring distances in lightyears and time in years for this example.) The second is when the twin in space reaches the destination, and the third is when the twins are together on Earth once more.

Imagine Tweedledee is the twin who stays on Earth, while Tweedledum is the twin launched into space.[9] Tweedledum is being launched on a round trip to Alpha Centauri, $4.37 lyr$ away, at the speed of $v = \frac{3}{5}c$.

Once again, this situation is most effectively examined using a Minkowski diagram. We begin with the axes representing Tweedledee's reference frame.

---

[9]The author hopes Mr. Carroll would not object to the use of his characters for educational purposes. Given the number of mathematical parables in *Alice's Adventures in Wonderland*, combined with the number of mathematical research papers published by Mr. Carroll under his birth name of Charles Lutwidge Dodgson, this seems somewhat likely.

ct

x

Tweedledum travels the $4.37 lyr$ to Alpha Centauri and back at speed $v = \frac{3}{5}c$:

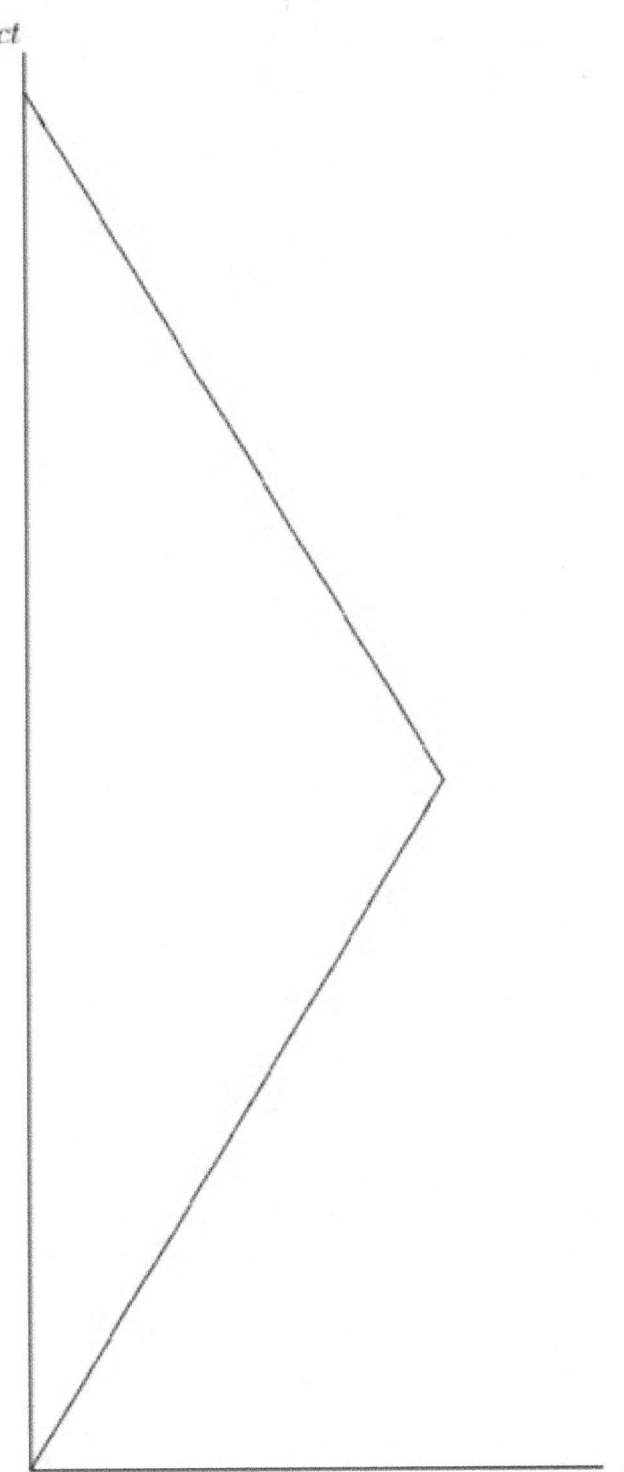

Recall that time dilation is an effect produced by the light speed delay in transmitting information. If Tweedledum watches Tweedledee, he can mark the years passed for Tweedledee as the light from the end of each year catches up to him. Marking those years, we see the following:

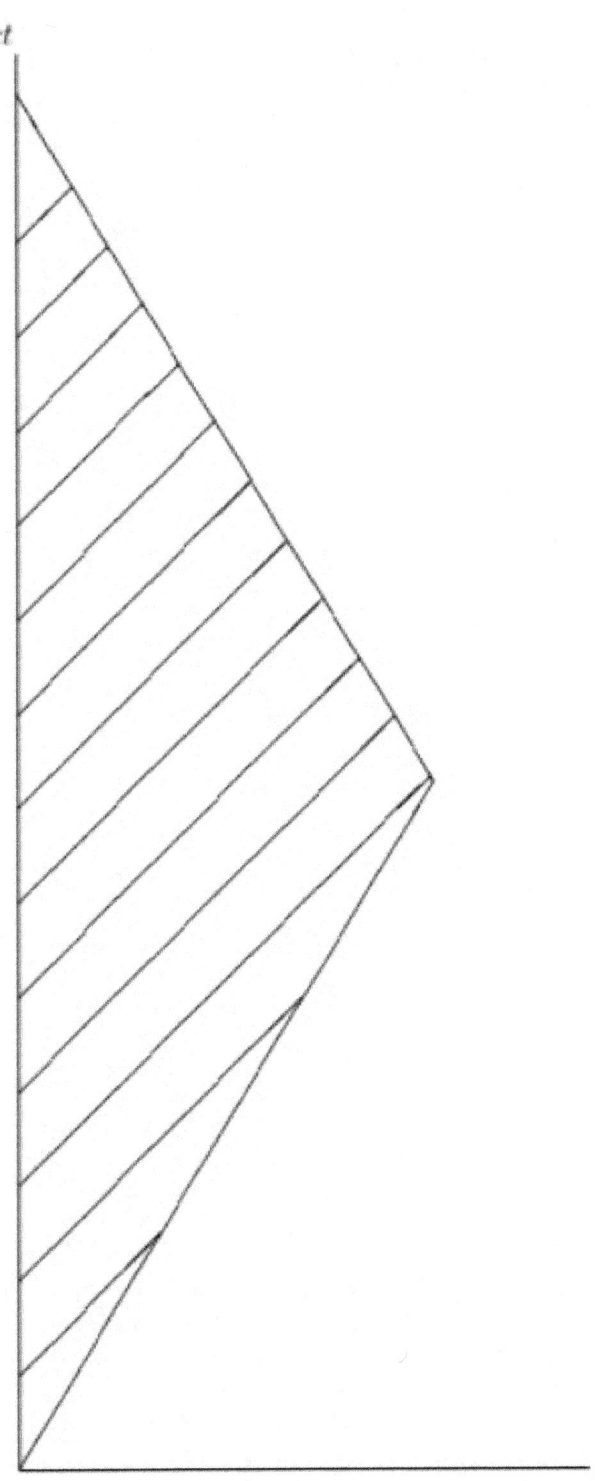

Note that time accelerates from Tweedledum's perspective on the return trip. In this trip, $v = -\frac{3}{5}c$, where the sign change represents the change in direction. If you look only at the section in which the twins move apart, each says the other is the younger twin, and each is justified in saying so within his own reference frame. Tweedledum sees Tweedledee age almost three years on the trip to Alpha Centauri, but he sees Tweedledee age nearly 12 years on the trip back. We now need to observe how Tweedledee ages.

If Tweedledee monitors Tweedledum's years in a similar fashion, we see a similar result. From Tweedledum's perspective, Alpha Centauri approaches him at $v = \frac{3}{5}c$, such that length contraction reduces the trip to $3.50lyr$, for a trip time of $5.83yr$ each way. Of course, we need to be sure we know where to mark the years on Tweedledum's path. The axes for the S' frame use different scales than those of the S frames. In this case, knowing that Tweedledum perceives the voyage as a $5.83yr$ trip each way, we divide Tweedledum's line into segments in proportion to the $5.83yr$ total, and mark the years at those points.

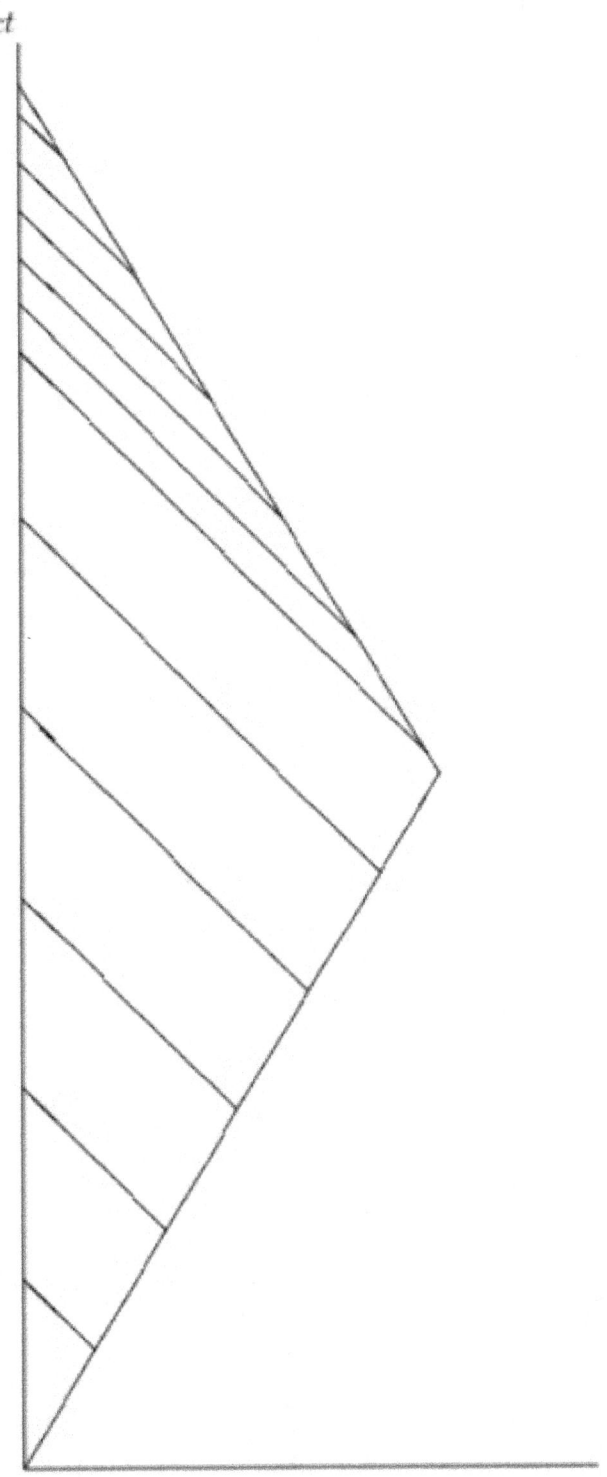

Thus, once reunited at the same point in spacetime after the trip, both twins agree that Tweedledee has aged $14.6yr$ and Tweedledum has aged $11.7yr$. Thus, Tweedledum's proper time is $\tau = 11.7yr$. The change in reference frames at Alpha Centauri for the return trip is enough to reconcile the paradox.

Chapter 5

Energy and Momentum

5.1 What Is Energy?

One of the quantities that physicists find most useful is energy. It is often difficult to define for new students, due in large part to the fact that it is an intangible quantity that you cannot see, touch, taste or smell. The formal definition is often "energy is the ability to do work," which seems clear and straightforward until one is asked to define "work" in the scientific sense.

An alternative definition of energy is "the ability to cause changes in motion or position." From a scientific perspective, this is equivalent to the first definition, but it reduces the terminology to that which is intuitive for new students. There are different types of energy in the world, though we will only deal with two of them in this series.

5.1.1 Kinetic Energy

The type of energy that is often the easiest for students to accept is kinetic energy, or energy of motion. If an object is in motion, its own position is constantly changing. It also has the potential to cause changes in motion to other bodies through collisions.

In the Newtonian view, kinetic energy can be calculated based solely on the mass and speed of an object. If two objects with different mass travel at the same speed, then the object with greater mass has greater kinetic energy. If two objects with the same mass travel at different speeds, then the one with the greater speed has the greater kinetic energy. Moreover, the speed of the object has significantly greater impact than the mass. (If you double an object's mass without changing its speed, you double its kinetic energy. If you double its speed, you more than double its kinetic energy.)

5.1.2   Potential Energy

The second major type of energy is potential energy. This is a little harder to see, as there is nothing visibly moving or altered in any way as a result of obtaining potential energy. This energy can be "released" and turned into kinetic energy in one form or another. For example, a book on a shelf has gravitational potential energy: the book might fall, gaining speed and kinetic energy as it approaches the floor. A book on a lower shelf has less potential energy, as it cannot fall as far and will be travelling at a lower speed when it hits the ground.

Potential energy is a consideration when there is some sort of outside field or force that can drive changes in motion, such as gravity. As such, further discussion will be saved for the general theory of relativity. The variables one needs to describe potential energy depend upon the source and nature of the outside force or field.

5.2   What Is Momentum?

Momentum is a term that is commonly used in day to day speech in a fashion that differs from its scientific meaning. When people talk about momentum in the colloquial sense, they typically mean something is in motion and hard to stop. This usage is closer to the word "inertia" than momentum. It is inertia that determines the resistance to accelerations and changes in motion. Momentum, in the Newtonian view, is the product of mass and velocity.[1] In Newton's day, it was referred to as the "quantity of motion." Momentum is always conserved in a collision, meaning that the total momentum that goes into a collision is the same as the total momentum that comes after. (Again, direction matters; two vehicles meeting head on can turn into a stationary wreck immediately after collision because momentum in one direction, say North, can be cancelled out by momentum in the opposite direction, South.)

5.3   How Are They Connected?

Kinetic energy and momentum can both be calculated using the same two pieces of information: the mass of the body in motion, and the velocity of the body in motion. Although they are not identical quantities,[2] the fact that they depend solely on the same two variables begs the question about whether or not the two are related.

5.3.1   The Newtonian View

In the Newtonian view, energy and momentum are connected through forces. When a force is applied to a object at rest, it accelerates to some velocity. The force applies over a period of time, during which the object travels some distance. Assuming the object is in space, so we don't have to worry about friction and other factors, allowing the entire force to apply to accelerating the object in the exact direction

---

[1]It uses velocity, and not speed, because direction matters.

[2]They cannot be the same: direction matters to momentum but not kinetic energy, which is a significant difference on the conceptual side.

of the force, then we can calculate both the energy and momentum using simple and similar math. We will also assume that the force is uniform, meaning it applies with the same strength and in the same direction the entire time it is applied. To calculate the energy of the object, one multiplies the force by the distance it has travelled. To calculate the momentum of the object, one multiplies the force by the time elapsed during the acceleration.[3]

5.3.2 The Relativistic View

In the relativistic view, there are a few changes that need to be made to these concepts. The first and foremost idea that needs to be explored is whether or not there is an upper limit to either quantity. After all, the universe has a speed limit in the speed of light. If these quantities depend on the speed of an object, then one would tend to think that there could be an upper limit to either or both in turn.

The relationship between the two quantities, as force applied along different directions (space or time), still applies in relativity. In that case, conceptually, we see that there should be no upper limit to either quantity: one can continually apply a force to an object, regardless of how much time elapses or distance is travelled. So, how do we obtain unlimited energy and momentum with a limited speed?

The answers lies in the fact that we have misidentified the variables involved. As we mentioned in lesson two, Newton originally recognized inertia as a key quantity, and then proposed that the mass and inertia of an object were one and the same quantity. Although he couldn't possibly have realized this at the time, mass and inertia are not, in fact, the same. The inertia of an object increases as its speed increases. This increase is almost imperceptible at the speeds that we see in our everyday experiences, but it is unmistakable at higher speeds. As an object gets closer to the speed of light, its inertia increases incredibly rapidly, ensuring that no amount of energy or momentum will carry the body above the speed of light. In several of the cases in which we used mass in Newtonian mechanics, we should have used inertia as Newton originally proposed. This concept was so counter intuitive, and the notion of equality between mass and inertia so widely accepted, that when this difficulty was first noticed physicists incorrectly assumed that the mass of an object was increasing, and the quantities were named as such. The mass was referred to as *rest mass* and the inertia was referred to as *relativistic mass*. Although the error in these terms is now understood, they were in use for so long that they still appear in much of the literature on the subject.

The next question pertains primarily to momentum: with momentum, direction matters, so what quantity appears as the time component of momentum? We now have four directions that need to be described when dealing with quantities that use direction at all. To discover this component, we begin with Newton's definition of momentum as the product of inertia and velocity. We take our four dimensional velocity and multiply it by the inertia of an object. What we find in the time component of the result is startling: it is the kinetic energy of the object, with an

---

[3]If the force is not uniform, some adjustments need to be made. Instead of simple multiplication, we use calculus to integrate the force with respect to the other variable.

additional term added: mc^2, where m is the mass of the object and c is the speed of light.

This was utterly shocking. An object at rest contained energy in the absence of outside fields or forces, and that energy E had the quantity mc^2. Never before had it occurred to anyone that mass could somehow store energy, and yet it seemed apparent that mass was some form of condensed, solidified energy. This revelation is the one that led to nuclear power. It was so unexpected that Einstein himself discussed it tangentially in his paper, deviating from the calculation he worked on to make a special point of describing it. It is also one of the primary examples of serendipitous discovery in the sciences: who would have believed that nuclear energy would have been discovered by examining momentum in a universe with a speed limit?

5.4   Newtonian Calculations

To better prepare the contrast between Newtonian calculations and relativistic calculations, we shall review the Newtonian formulae of mechanics and show how they are derived.

5.4.1   Force

The expression for force is one of the few *axiomatic* formulae in classical mechanics. As an axiomatic formula, it has no derivation; it is simply believed to be true. Newton proposed force as the product of inertia and acceleration. He soon proposed that an object's mass was identical to its inertia, resulting in the formula

$$\mathbf{F} = m\mathbf{a} = m\frac{d\mathbf{v}}{dt} = m\frac{d^2\mathbf{x}}{dt^2}$$

Much of the rest of mechanics are drawn from these formulae.

5.4.2   Momentum

Momentum was proposed through the relationship

$$\mathbf{F} = \frac{d\mathbf{p}}{dt}$$

In other words, force causes an object to change momentum over time. In cases in which the mass of the accelerated object remains constant, this reduces simply though an integral to

$$\mathbf{p} = \int d\mathbf{p} = \int \mathbf{F}dt = \int m\frac{d\mathbf{v}}{dt}dt = \int md\mathbf{v} = m\mathbf{v}$$

which is the version people are most accustomed to seeing. In cases with variable mass, such as any self-propelled, fuel burning object, the final expression becomes more complicated, but it is still based on integrating the applied force over time.

5.4.3 Energy

Energy E is defined scientifically as the ability to do work W, or to change the state of motion of an object. There are two types of energy, primarily defined through the concepts of force and work. The two forms are kinetic and potential energy, and kinetic energy is the easiest to derive. Both are based on the idea that

$$W = \Delta E = \int \mathbf{F} \cdot d\mathbf{x}$$

In other words, applying a force in the direction of motion of an object changes its state of motion and does work, changing its energy.

Kinetic Energy

The kinetic energy derivation is the simplest to do in many ways. Again, the familiar form is based on the idea that the mass (or inertia) of an object remains constant during the motion of the body. If an object starts with zero energy and is accelerated to velocity v, then the kinetic energy E_k is given by

$$E_k = \int \mathbf{F} \cdot d\mathbf{x} = \int m\frac{d\mathbf{v}}{dt} \cdot d\mathbf{x} = \int m d\mathbf{v} \cdot \frac{d\mathbf{x}}{dt} = \int m d\mathbf{v} \cdot \mathbf{v} = \frac{1}{2}mv^2$$

which is the familiar form.

Potential Energy

Potential energy is a bit more difficult to recognize. Imagine you ride an elevator up three floors in a building. We know energy is being applied, even through the middle portion of the ride which is at a constant speed, as the elevator car continues to climb and consume electrical power. The net change in speed, however, is zero: you were motionless before the elevator started to climb, and you were motionless when you reached the top. This is a sign that your potential energy has changed: were you to step into the elevator shaft when no elevator was present, gravity would quickly cause you to accelerate to ground level. Similar situations can be developed when charged particles are close to each other, or magnets are close to each other, and so forth. Any time one might experience a force of some kind, there is a potential involved.

One of the most common potentials we face is gravity. In the case of first exposure to this force, we tend to work with problems at or near the surface of the planet Earth, where the acceleration due to gravity is given by $g = 9.81\frac{m}{s^2}$. in the downward direction.[4] This value of g remains fairly constant near the planet's surface. Because of this near constancy, we can calculate the potential energy of an object a height h from the planet's surface by examining the action of the force of gravity as the object falls.

---

[4]In fact, when one tries to come up with a scientific definition of the word "down," one is forced to define it as the direction that the force of gravity points.

$$\Delta E_p = \int \mathbf{F} \cdot d\mathbf{x} = \int m\mathbf{g} \cdot d\mathbf{x} = \int_h^{0m} mgdz = -mgh$$

This is almost the familiar form. The negative sign appears because this calculation represents the change in potential energy as an object drops *from* a height h to ground level; the object has lost mgh worth of potential energy. By convention, ground level is taken to be the point at which potential energy is zero, so this means the object had a positive mgh worth of energy when it was still at height h. Thus,

$$E_p = mgh$$

is the familiar expression for an object at height h above the Earth's surface. This works for the balls dropped off bridges and buildings in most introductory problems, but doesn't work very well for objects in orbit. With distances that great, g cannot be treated as a constant. In those cases, we work with the full gravitational force expression

$$F_g = \frac{Gm_1m_2}{r^2}$$

where m_1 and m_2 are the masses of the two objects involved, $G = 6.67 \times 10^{-11} \frac{Nm^2}{s^2}$ is Newton's gravitational constant, and r is the distance between the centres of mass of the two objects. Integrating over r results in the expression

$$E_p = \int \mathbf{F}_g \cdot d\mathbf{r} = \int \frac{Gm_1m_2}{r^2}dr = -\frac{Gm_1m_2}{r}$$

The negative sign appears because of a convention that results from something known as *gauge freedom*. When doing these experiments, we often decide arbitrarily where the point of zero height is in our apparatus. When dropping things in a lab, it is convenient to choose the floor, the lab bench, or whatever other target we are using as the point of zero energy. We do this no matter which floor of the building we are on, knowing full well that the object we drop could continue to build up kinetic energy if we were to drop it down the stairs or into a floor drain instead. We have the freedom to set the "zero" of our gauge to whatever level is most convenient; the integral definition we use emphasizes the fact that it is the change in position that is important, and not the position itself. So, where is our zero energy point when dealing with situations too big to fit in a lab room, such as an orbiting satellite? We set the zero energy point at infinity, making potential energies *negative* as one object accelerates and approaches the other object.[5] This also means that an object that starts at rest at infinity and falls towards the other body arrives with kinetic

---

[5]Technically, both accelerate towards each other, but in most real life problems, we typically treat the planet or star as an object at a fixed point in space and the satellite as moving. The motion of the planet or star is generally too small to be measured or relevant.

and potential energies that still add up to zero, so that $E_k = -E_p$, which makes other calculations convenient.

Of course, further expressions could be developed for other forces. For example, a complete electromagnetic potential could be developed by integrating the Lorentz force law:

$$E_p = \int q(\mathbf{E} + \mathbf{v} \times \mathbf{B}) \cdot d\mathbf{x}$$

but these are not directly related to what is coming next.

5.5 Relativistic Calculations

5.5.1 Time Derivatives and Invariants

In order to work effectively in relativity, we need to ensure that our four vectors are Lorentz invariant. We have established for our position four-vector already, but we still need to confirm this for our four dimensional velocity and its timelike component.

Let us begin with the instinctive option:

$$\vec{u}' = \frac{d\vec{x}'}{dt'} = \frac{d}{dt'} \begin{pmatrix} ct' \\ x' \\ y' \\ z' \end{pmatrix} = \frac{d}{dt'} \begin{pmatrix} \gamma \left(ct - \frac{v}{c}x \right) \\ \gamma \left(x - vt \right) \\ y' \\ z' \end{pmatrix} = \begin{pmatrix} c \\ \frac{u_x - v}{1 - \frac{u_x v}{c^2}} \\ u_y \\ u_z \end{pmatrix}$$

Comparing this to

$$\vec{u} = \frac{d\vec{x}}{dt} = \begin{pmatrix} c \\ u_x \\ u_y \\ u_z \end{pmatrix}$$

we see that this is only invariant when

$$u_x = \frac{u_x - v}{1 - \frac{u_x v}{c^2}}$$

which can be manipulated to the point

$$vu_x^2 = vc^2$$

This is true in only two situations. If $v = 0$, this is true, but hardly useful; that means S and S' are the same reference frame. If $v \neq 0$, then we have $u_x = \pm c$, which is hardly useful in general situations. Thus, our instinct for an invariant four-vector fails us, and we need to use another option.

The problem lies in the definition's right hand side $\frac{d\vec{x}'}{dt'}$. The numerator is the differential of an invariant quantity, while the denominator is not. To correct this, we need to differentiate with respect to another invariant quantity. Thankfully, this was introduced in the previous chapter, though it did not seem that significant at the time.

With most time derivatives, one must differentiate with respect to the proper time, and not the observer's time, when performing calculations. In essence, we are treating the S' frame as the reference frame in which the moving body is at rest. In short, we are setting $\vec{v} = \vec{u}$ and using this as our reference. With this definition,

$$c^2 \tau^2 = -c^2 t^2 + x^2 + y^2 + z^2$$

is an invariant quantity. When the object is at rest, we find

$$d\tau = dt \sqrt{1 - \frac{u^2}{c^2}} = \frac{dt}{\gamma(u)}$$

which is, essentially, our time dilation equation for $u = v$, with γ treated as a function of u.

This leads us to the four vector definition

$$\vec{u} = \frac{d\vec{x}}{d\tau} = \frac{d\vec{x}}{dt}\frac{dt}{d\tau} = \gamma(u) \begin{pmatrix} c \\ u_x \\ u_y \\ u_z \end{pmatrix}$$

Now we calculate $\vec{u} \cdot \vec{u}$ to test invariance, noting that $v = u$:

$$\vec{u} \cdot \vec{u} = \gamma \begin{pmatrix} c & u_x & u_y & u_z \end{pmatrix} \begin{pmatrix} -1 & 0 & 0 & 0 \\ 0 & 1 & 0 & 0 \\ 0 & 0 & 1 & 0 \\ 0 & 0 & 0 & 1 \end{pmatrix} \gamma \begin{pmatrix} c \\ u_x \\ u_y \\ u_z \end{pmatrix}$$

$$= \gamma^2 \left(-c^2 + u_x^2 + u_y^2 + u_z^2 \right)$$

$$= \frac{-c^2 + u^2}{1 - \frac{u^2}{c^2}}$$

$$= -c^2 \frac{1 - \frac{u^2}{c^2}}{1 - \frac{u^2}{c^2}}$$

$$= -c^2$$

The final result for this quantity is independent of any information about $\vec{u}$ in any way. Thus, invariance results.

When acceleration is involved, things are sadly more complicated. Invariance is true when dealing with an inertial reference frame. We cannot use the S' frame as the rest frame for an accelerated object, as this is *not* an inertial reference frame. The very nature of acceleration shows this to be the case, making any hope of invariance rather slim. Thus, we define the acceleration as

$$\vec{a} = \frac{d\vec{u}}{d\tau} = \frac{d^2\vec{x}}{d\tau^2}$$

with little hope of an invariant quantity. In fact,

$$\vec{a} = \frac{d\vec{u}}{d\tau} = \frac{d\vec{u}}{dt}\frac{dt}{d\tau} = \gamma\frac{d\vec{u}}{dt} = \gamma^2 \begin{pmatrix} \frac{ua\gamma}{c^2} \\ u_x\frac{ua\gamma^2}{c^2} + a_x \\ u_y\frac{ua\gamma^2}{c^2} + a_y \\ u_z\frac{ua\gamma^2}{c^2} + a_z \end{pmatrix}$$

If we look specifically at the rest frame of the object in the exact moment the acceleration is first applied, we find that $u = 0$ and $a = \alpha$, which we can refer to as the proper acceleration, or the acceleration at measured by the object at rest. In this frame,

$$\vec{a} = \begin{pmatrix} 0 \\ \alpha_x \\ \alpha_y \\ \alpha_z \end{pmatrix}$$

As it turns out, this leads to an invariant quantity:

$$\vec{a} \cdot \vec{a} = \vec{a}' \cdot \vec{a}' = \alpha^2$$

Thus, even though it seemed unlikely that we'd reach an invariant, we have, further justifying our use of $d\tau$ as the preferred time differential for relativity.

5.5.2 Force

The relativistic force is relatively easy to compute. As with Newtonian mechanics, it is the product of mass and acceleration. We define the four-vector force as the four-vector acceleration multiplied by the mass of the object:

$$\vec{F} = m_0\vec{a} = m_0\frac{d\vec{u}}{d\tau}$$

The subscript on the mass m_0 is there because we will soon find that mass and inertia are not identical as was believed in Newton's day. This form also assumes that mass is constant, which is not necessarily the case. The invariance of this form is relatively simple to establish: $\vec{F} \cdot \vec{F} = m_0^2\vec{a} \cdot \vec{a} = m_0^2\alpha^2$, which is invariant.

In the case in which the mass is variable, the new form becomes

$$\vec{F} = \frac{d(m_0\vec{u})}{d\tau}$$

5.5.3 Momentum

Now that we have

$$\vec{F} = \frac{d(m_0\vec{u})}{d\tau}$$

for our force, we can build our momentum in an analogous fashion to the way we did in the Newtonian view. In that case,

$$\vec{F} = \frac{d(m_0\vec{u})}{d\tau} = \frac{d\vec{p}}{d\tau}$$

Thus,

$$\vec{p} = \int \vec{F} d\tau = \int \frac{d(m_0\vec{u})}{d\tau} d\tau = int d(m_0\vec{u}) = m_0\vec{u}$$

which probably would have been our instinctive choice to begin with. Now, doing this with algebra is all well and good, but what do the components mean? The spatial components of $\vec{p}$ are easy enough to interpret, but now we have a time component as well. What is the natural analogue to this component?

Let us examine this four-vector explicitly.

$$\vec{p} = \begin{pmatrix} \gamma m_0 c \\ \gamma m_0 u_x \\ \gamma m_0 u_y \\ \gamma m_0 u_z \end{pmatrix}$$

When the object is at rest and $u = 0 m/s$, we are left with

$$\vec{p} = \begin{pmatrix} m_0 c \\ 0 \\ 0 \\ 0 \end{pmatrix}$$

We see that the time component is the mass of the object multiplied by the speed of light, but what does that *mean*? This is some sort of "momentum through time," but what is a momentum through time? The quantity must be important in some way, as we find that the invariant length of our momentum vector is closely related:

$$\vec{p} \cdot \vec{p} = -m_0^2 c^2$$

So, this combination of mass and the speed of light must be important, but we need to see why that is.

5.5.4 Energy

In the three dimensional world, we knew that

$$\mathbf{F} = \frac{d\mathbf{p}}{dt}$$

and

$$F = \frac{dE}{dx}$$

If the same relationships hold in relativistic mechanics, then we would have something analogous to

$$\frac{dp}{dt} = \frac{dE}{dx}$$

to relate the two quantities. Dealing with this in four dimensional space, we have

$$E = \int d\vec{p} \cdot \frac{d\vec{x}}{d\tau} = \int \vec{u} \cdot d\vec{p}$$

In the case when $u = 0 m/s$, we are left only with the time components of these vectors:

$$E = m_0 c^2$$

which may be the single most famous equation in the history of science. Furthermore, this is a natural occurrence of the time component of momentum. Our "momentum through time" is nothing more than the energy of the object divided by the speed of light. Even more surprisingly, objects at rest with no potential energy still have energy by virtue of mass! Performing the calculation in the general form gives us that

$$E = \gamma m_0 c^2 = m c^2$$

where

$$m = \gamma m_0$$

is the inertia of the object in motion. Thus, energy and inertia are revealed to be closely related aspects of the same quantity. It was this discovery, coupled with quantum mechanics, that led to the development of nuclear technology.

Kinetic Energy

If the energy of an object at rest is

$$E = m_0 c^2$$

and the total energy of an object in motion (with no potential energy) is given by

$$E = \gamma m_0 c^2$$

then it stands to reason that the kinetic energy of an object is the difference between these two:

$$E_k = \gamma m_0 c^2 - m_0 c^2 = m_0 c^2 \left(\frac{1}{\sqrt{1 - \left(\frac{u}{c}\right)^2}} - 1 \right)$$

This looks nothing at all like the familiar

$$E_k = \frac{1}{2} m v^2$$

of the Newtonian world that works so well in the high school lab. How can it possibly be correct?

Those who have taken calculus may be familiar with Taylor series expansions. The basic concept is that any continuous function with continuous derivatives can be approximated by a (possibly infinite) polynomial. If one takes the relativistic kinetic equation above and applies this technique, one finds that the infinite approximating polynomial begins

$$E_k = \frac{1}{2}m_0 u^2 + \frac{3}{8}m_0 \frac{u^4}{c^2} + \dots$$

The first time is the formula we are used to. The next term is one that is divided by c^2, which is no small number. In fact, each term is of the form

$$\frac{p}{q}m_0 \frac{u^{k+2}}{c^k}$$

where k, p and q are all whole numbers with $p < q$. As a result, when $u \ll c$, the relativistic nature of the inertia is difficult to detect, and the Newtonian form is an excellent approximation. It is also worth noting that the relativistic kinetic energy formula derived here is one that allows for unlimited energy, but contains an implicit speed limit.

Potential Energy

When objects have a significant amount of potential energy, they tend to experience a variety of accelerations. Most potential energy formulae are virtually unchanged (aside from the replacement of mass with inertia). These will be developed more fully in chapters seven through nine.

Chapter 6

Electricity and Magnetism

6.1 Electricity and Magnetism the Newtonian Way

We have already seen how the laws of electricity and magnetism led to the rise of relativity. Let us examine how relativity alters them more closely.

Both electrical and magnetic phenomena have been studied for thousands of years. It was not until 1820, however, that the scientific community at large recognized how closely connected the two are. Hans Christian Oersted noticed (which giving a lecture to students) that electric currents deflect the needle of a compass. Further studies were able to quantify the relationship, explicitly defining and mathematically defining the procedure. It was clear that an electrically charged object in motion produces a magnetic field that is circular in shape and surrounds the particle at a right angle to the direction of motion. There was a problem, though: there was no theoretical reason to connect the two interactions. The two types of phenomena were kludged together mathematically because the connection was experimentally proven, and not because of any intrinsic motivation to tie the two together. It was comparable to rain and rainbows: people recognized that rainbows appeared after rainfalls so early on that even the word "rainbow" includes the word "rain," but it would be centuries before the optics involved were understood well enough to explain precisely why rainbows appeared.

The theoretical descriptions of electrical and magnetic phenomena was completed in the late 1860s, and James Clerk Maxwell collated the results into what are now known as Maxwell's equations. These described all known electrical and magnetic phenomena, detailing the connections between the fields without explaining why they are connected, with particular focus and attention on how the electric and magnetic fields change and evolve through space and time.[1] Just as ancient people could predict the appearance of rainbows, people in the 1900s could predict the magnetic fields produced by electrically charged objects to great effect, but the connection was arbitrary. This was still a young science, so when it was shown

---

[1] Remember, to James Clerk Maxwell and his contemporaries, space and time were two completely distinct quantities.

to be inconsistent with Newtonian physics, it was assumed that the fault lay with electricity and magnetism. As both theories were consistent with the experimental data available at the time, this assumption seemed reasonable. As we learned in earlier lessons, it was actually the Newtonian mechanics that were at fault.

6.2   Electricity and Magnetism the Relativistic Way

The work of Lorentz and Einstein combined to solve the problems with electricity and magnetism in multiple frames of reference. Furthermore, they showed that electromagnetic theory did not need to be altered in any special way; the framework established by Maxwell's equations was completely consistent with electricity and magnetism. There were, however, other significant benefits to the relativistic framework that were worth exploring.

So far, when we take a relativistic viewpoint of quantities, we have found a common pattern: quantities that depend on direction get a fourth quantity attached which serves as the "time direction" component of the quantity. In all such cases, the fourth quantity was independent of direction. In the case of electrical fields, the prime candidate for such a fourth quantity is the magnetic field, but that *also* depends on direction. It can't fit easily into that fourth slot. This seems to be a stumbling block on the path to connecting electricity and magnetism on the theoretical side.

Undeterred, we go back to what we learned in lesson four: although no simple rotation can move something from a spatial direction into the time direction, one can formulate a rotation based on acceleration that does the same job. This becomes the key to connecting the two phenomena.

Imagine, if you will, a long, electrically charge rod that is at rest. In the classical view, accelerating this rod along its axis[2] causes no change in the electrical field, but produces a sudden and inexplicable magnetic field. If we had two such rods lying parallel and moving in identical directions at identical speeds, the magnetic force created would be an attractive force.

We now know that such a motion will increase the electrical force: as the rods accelerate, their lengths contract. With the same amount of charge packed into a shorter distance, we see the electrical charge density increase, which causes an increase in the electrostatic attraction. By carefully calculating the result of "rotating" the electric field to a higher velocity using the math that describes our relativistic world, we find that the action of the classical magnetic field is identical to the action of this "extra" electric field produced by relativity! This math is independent of the situation, so it doesn't matter if our moving electrical charge is a rod, disc, point, etc. This is true for any shape that moves while carrying an electrical charge.

This was something of a holy grail to physicists: the elusive theoretical connection between the electric and magnetic fields that didn't exist in the classical view became an immediate and natural consequence of relativistic dynamics. The magnetic fields

---

[2]By "along its axis," we mean that if the rod lies in the North/South direction, then we accelerate it either North or South, and not East, West, up or down.

observed around active circuits resulted directly from the motion of charged particles under the rules of relativity. The mathematical object used to combine the two is more complicated than a four directional vector, but it is just as valid: the fields were unified into a single phenomenon, and Maxwell's set of four[3] equations simplified to a single equation. This served as yet another triumph of the special theory of relativity. Despite a rocky start, the scientific community had finally accepted special relativity as the theory that explained reality better than any other theory. It is now known that any errors in relativity are small, as the theory has been tested with incredible experimental rigour. Still, special relativity is limited to points of view that do not experience forces or accelerations. Most of our work is done in the presence of a gravitational force, and many real world observers need to change their direction to do their jobs. The special theory of relativity was inadequate for these tasks, and would need to be expanded into the general theory of relativity, which is the subject of our final lessons.

6.3 Maxwell

James Clerk Maxwell compiled all that was known about electricity and magnetism into four equations. Those equations are as follows:

$$\nabla \cdot \mathbf{E} = \frac{1}{\epsilon_0}\rho \tag{6.1}$$

$$\nabla \times \mathbf{E} = -\frac{\partial \mathbf{B}}{\partial t} \tag{6.2}$$

$$\nabla \cdot \mathbf{B} = 0 \tag{6.3}$$

$$\nabla \times \mathbf{B} = \mu_0 \mathbf{J} + \mu_0 \epsilon_0 \frac{\partial \mathbf{E}}{\partial t} \tag{6.4}$$

where

$$\nabla = \begin{bmatrix} \frac{\partial}{\partial x} \\ \frac{\partial}{\partial y} \\ \frac{\partial}{\partial z} \end{bmatrix}$$

is the usual differential operator, $\mathbf{E}$ is the electric field, ϵ_0 is the electric permittivity of space, ρ is the charge density, $\mathbf{B}$ is the magnetic field, t is time, μ_0 is the magnetic permeability of free space and $\mathbf{J}$ is the current density. As Lorentz already showed, these equations were consistent with relativity from the start.

6.4 General Mathematical Tools

With the electric and magnetic fields intrinsically connected, a simple four vector will not be enough to describe them. There are at least six components of these fields to deal with; three spatial components of the electric field, and three spatial

---

[3]Well, technically, Maxwell did his work before physicists adopted vectors as mathematical tools, so his "four" equations were originally published as twelve equations. In a roundabout sort of way, this also explains why the symbols for so many electromagnetic variables are counter intuitive.

components of the magnetic field. Before we can effectively formulate what this new tensor would look like, we need to introduce some mathematical shorthand tools.

6.4.1   The Levi-Civita Symbol

The Levi-Civita symbol is one that is quite useful in vector analysis. It has a minimum of three indices, and takes one of three values:

$$
\epsilon_{\mu\nu\alpha} = \begin{cases} 1 & \mu\nu\alpha \text{ for am even permutation of 1, 2, and 3. (e.g. 123, 231, 312)} \\ -1 & \mu\nu\alpha \text{ form an odd permutation of 1, 2, and 3. (e.g. 132, 321, 213)} \\ 0 & \text{otherwise (e.g. any index repeated)} \end{cases}
$$

The Levi-Civita symbol makes cross products a bit easier to manage. Under the normal definition, the cross product $\mathbf{A} \times \mathbf{B} = \mathbf{C}$ is defined as the determinant of the matrix-type object

$$
\begin{vmatrix} \hat{e}_x & \hat{e}_y & \hat{e}_z \\ A_x & A_y & A_z \\ B_x & B_y & B_z \end{vmatrix} = \begin{bmatrix} A_y B_z - A_z B_y \\ A_z B_x - A_x B_z \\ A_x B_y - A_y B_x \end{bmatrix}
$$

This has two big disadvantages. From the mathematical perspective, it is putting vectors and scalars together into a "matrix" and taking the determinant, but a matrix requires all of its entries to be of the same type. Thus, it's not technically a permissible mathematical object. From the educational perspective, it is often useful to teach cross products before linear algebra has been taught, resulting in a haphazard teaching method that only allows component wise work when the vectors $\mathbf{A}$ and $\mathbf{B}$ are perfectly perpendicular. There is, however, a certain regularity and a pattern to these subscripts. We can redefine the cross product components as follows, given that $\mathbf{A} \times \mathbf{B} = \mathbf{C}$:

$$
C_i = \sum_j \sum_k \epsilon_{ijk} A_j B_k
$$

To calculate C_x, for example, we calculate

$$
\begin{aligned}
C_x &= \epsilon_{xxx} A_x B_x + \epsilon_{xxy} A_x B_y + \epsilon_{xxz} A_x B_z \\
&+ \epsilon_{xyx} A_y B_x + \epsilon_{xyy} A_y B_y + \epsilon_{xyz} A_y B_z \\
&+ \epsilon_{xzx} A_z B_x + \epsilon_{xzy} A_z B_y + \epsilon_{xzz} A_z B_z \\
C_x &= 0 + 0 + 0 \\
&+ 0 + 0 + A_y B_z \\
&+ 0 - A_z B_y + 0 \\
C_x &= A_y B_z - A_z B_y
\end{aligned}
$$

There are a few more terms involved at first glance, but with a bit of practice this is highly efficient, as the mathematician/physicist/student in question chooses not to write down the terms with repeated indices in the first place. The Levi-Civita symbol easily extends to four dimensions as well, using indices 0, 1, 2, and 3.

6.4.2 Derivatives with Tensors

When we begin working in electromagnetism, taking derivatives will be an unavoidable situation. We need to be careful, however, since we have both four-vector components and one form components. The four-vector components are unmodified by the metric, but the one form components are, so taking the derivative with respect to each provides different results. We want to differentiate four-vectors with respect to four-vector components and one forms with respect to one form components. Algebraically, we represent this with commas in our subscripts and superscripts.

We define a partial derivative with respect to a four-vector component as

$$\frac{\partial A^\mu}{\partial x^\nu} = A^\mu{}_{,\nu}$$

Notice that a raised exponent in the denominator becomes a lowered exponent in the numerator. This facilitates the use of Einstein summation convention. Notice also the comma that is in place to represent the partial derivative.

We have already made use of the ∇ operator for taking derivatives in three dimensions, as defined above. The three sided shape is useful for a three dimensional differential operator. The convention for the four dimensional differential operator is the natural extension of this:

$$\vec{\Box} = \begin{pmatrix} \frac{\partial}{\partial x^0} \\ \frac{\partial}{\partial x^1} \\ \frac{\partial}{\partial x^2} \\ \frac{\partial}{\partial x^3} \end{pmatrix}$$

Thus,

$$\Box^2 = \Box^\mu \Box_\mu = -\frac{\partial^2}{c^2 \partial t^2} + \frac{\partial^2}{\partial x^2} + \frac{\partial^2}{\partial y^2} + \frac{\partial^2}{\partial z^2}$$

where $\Box^\mu$ is the derivative with respect to *one form* components and $\Box_\mu$ is the derivative with respect to *four vector* components. This is somewhat counter intuitive, but it is so defined to facilitate the use of Einstein summation convention in cases such as

$$\vec{\Box} \cdot \vec{A} = \Box_\mu A^\mu$$

6.4.3 Symmetric and Antisymmetric Tensors

Imagine a 4×4 tensor. Generally speaking, this tensor will have 16 possible components. There are, however, two circumstances in which these can be simplified to either 10 or 6 components.

Symmetric Tensors

Our metric

$$g^{\mu\nu} = \begin{pmatrix} -1 & 0 & 0 & 0 \\ 0 & 1 & 0 & 0 \\ 0 & 0 & 1 & 0 \\ 0 & 0 & 0 & 1 \end{pmatrix}$$

and our Lorentz transformation tensor

$$\Lambda^{\mu'}{}_{\nu} = \begin{pmatrix} \gamma & -\frac{v}{c}\gamma & 0 & 0 \\ -\frac{v}{c}\gamma & \gamma & 0 & 0 \\ 0 & 0 & 1 & 0 \\ 0 & 0 & 0 & 1 \end{pmatrix}$$

are both symmetric tensors. Symmetric tensors are defined by the property

$$S^{\mu\nu} = S^{\nu\mu}$$

indicating a symmetry about the diagonal line that starts in the upper left corner of the tensor and continues to the lower right corner of the tensor. In this case, there are only 10 independent entries. Once S^{01} has been determined, S^{10} follows immediately. In the case of an antisymmetric tensor,

$$A^{\mu\nu} = -A^{\nu\mu}$$

which reduces the situation to only 6 independent entries: A^{01} still determines A^{10} and the other five entries on that side of the diagonal. The entries on the diagonal itself must conform to

$$A^{\mu\mu} = -A^{\mu\mu} = 0$$

so that the diagonal of every antisymmetric tensor is identically 0. This is our best bet for the shape of the object that will be our electromagnetic field tensor, containing information about the three electric field components and the three magnetic field components.

6.5   The Electromagnetic Field Tensor

The electromagnetic field tensor cannot be explicitly derived from scratch, simply because there is more than one way to formulate it. Instead of deriving them, the two forms will be displayed and shown to behave as required.

The most common form of the electromagnetic field tensor is

$$F^{\mu\nu} = \begin{pmatrix} 0 & \frac{E_x}{c} & \frac{E_y}{c} & \frac{E_z}{c} \\ -\frac{E_x}{c} & 0 & B_z & -B_y \\ -\frac{E_y}{c} & -B_z & 0 & B_x \\ -\frac{E_z}{c} & B_y & -B_x & 0 \end{pmatrix}$$

The alternate form is known as the dual tensor, and it is

$$G^{\mu\nu} = \begin{pmatrix} 0 & B_x & B_y & B_z \\ -B_x & 0 & -\frac{E_z}{c} & \frac{E_y}{c} \\ -B_y & \frac{E_z}{c} & 0 & -\frac{E_x}{c} \\ -B_z & -\frac{E_y}{c} & \frac{E_x}{c} & 0 \end{pmatrix}$$

The structures are clearly similar between the two forms.

To show that this is useful, we must prove that it transforms and reacts to Lorentz transformations as expected. In tensor notation, the transformation would be written

$$F^{\mu'\nu'} = \Lambda^{\mu'}{}_{\mu}\Lambda^{\nu'}{}_{\nu}F^{\mu\nu}$$

which requires two instances of the Lorentz transformation tensor to transform the two indices. Let us look at a simplified example: we look at a situation in which a charged particle is at rest in frame S, producing no magnetic field, and is boosted into the S' frame. If we use only one Lorentz transformation tensor we have

$$\begin{pmatrix} \gamma & -\frac{v}{c}\gamma & 0 & 0 \\ -\frac{v}{c}\gamma & \gamma & 0 & 0 \\ 0 & 0 & 1 & 0 \\ 0 & 0 & 0 & 1 \end{pmatrix} \begin{pmatrix} 0 & \frac{E_x}{c} & \frac{E_y}{c} & \frac{E_z}{c} \\ -\frac{E_x}{c} & 0 & 0 & 0 \\ -\frac{E_y}{c} & 0 & 0 & 0 \\ -\frac{E_z}{c} & 0 & 0 & 0 \end{pmatrix}$$
$$= \begin{pmatrix} \frac{v\gamma E_x}{c^2} & \frac{\gamma E_x}{c} & \frac{\gamma E_y}{c} & \frac{\gamma E_z}{c} \\ -\frac{\gamma E_x}{c} & -\frac{v\gamma E_x}{c^2} & -\frac{v\gamma E_y}{c^2} & -\frac{v\gamma E_z}{c^2} \\ -\frac{\gamma E_y}{c} & 0 & 0 & 0 \\ -\frac{\gamma E_z}{c} & 0 & 0 & 0 \end{pmatrix}$$

which lacks the antisymmetric property we require. This is because we have only transformed one of the two indices; we have transformed columns but not rows, so that the components have not been transformed completely. Transforming the rows (by placing the matrix Λ^T on the right and multiplying) results in

$$\begin{pmatrix} \frac{v\gamma E_x}{c^2} & \frac{\gamma E_x}{c} & \frac{\gamma E_y}{c} & \frac{\gamma E_z}{c} \\ -\frac{\gamma E_x}{c} & -\frac{v\gamma E_x}{c^2} & -\frac{v\gamma E_y}{c^2} & -\frac{v\gamma E_z}{c^2} \\ -\frac{\gamma E_y}{c} & 0 & 0 & 0 \\ -\frac{\gamma E_z}{c} & 0 & 0 & 0 \end{pmatrix} \begin{pmatrix} \gamma & -\frac{v}{c}\gamma & 0 & 0 \\ -\frac{v}{c}\gamma & \gamma & 0 & 0 \\ 0 & 0 & 1 & 0 \\ 0 & 0 & 0 & 1 \end{pmatrix}$$
$$= \begin{pmatrix} 0 & \frac{E_x}{c} & \frac{\gamma E_y}{c} & \frac{\gamma E_z}{c} \\ -\frac{E_x}{c} & 0 & -\frac{v\gamma E_y}{c^2} & -\frac{v\gamma E_z}{c^2} \\ -\frac{\gamma E_y}{c} & \frac{v\gamma E_y}{c^2} & 0 & 0 \\ -\frac{\gamma E_z}{c} & \frac{v\gamma E_z}{c^2} & 0 & 0 \end{pmatrix}$$

which is the antisymmetric tensor with the correctly transformed components:

$$\mathbf{E'} = \begin{bmatrix} E_x \\ \gamma E_y \\ \gamma E_z \end{bmatrix}$$

and

$$\mathbf{B}' = \begin{bmatrix} 0 \\ \frac{v\gamma E_z}{c^2} \\ -\frac{v\gamma E_y}{c^2} \end{bmatrix}$$

6.6   Maxwell Redux

In their relativistic form, Maxwell's equations in free space (in which ρ and $\mathbf{J}$ are both zero) can be reduced to a single equation by using the electromagnetic field tensor and its derivatives.

We begin by examining the derivatives of the electromagnetic field tensor. Starting with the derivatives of the first row, we get

$$\begin{aligned} F^{0\nu}{}_{,\nu} &= \frac{\partial F^{00}}{\partial x^0} + \frac{\partial F^{01}}{\partial x^1} + \frac{\partial F^{02}}{\partial x^2} + \frac{\partial F^{03}}{\partial x^3} \\ &= \frac{1}{c}\left(\frac{\partial E_x}{\partial x} + \frac{\partial E_y}{\partial y} + \frac{\partial E_z}{\partial z} \right) \\ &= \nabla \cdot \mathbf{E} = \frac{1}{\epsilon_0}\rho \end{aligned}$$

where we have the first of Maxwell's equations, presented above as Equation (6.1), in the final step.

Examination of the second row gives us

$$\begin{aligned} F^{1\nu}{}_{,\nu} &= \frac{\partial F^{10}}{\partial x^0} + \frac{\partial F^{11}}{\partial x^1} + \frac{\partial F^{12}}{\partial x^2} + \frac{\partial F^{13}}{\partial x^3} \\ &= -\frac{1}{c^2}\frac{\partial E_x}{\partial t} + 0 + \frac{\partial B_z}{\partial y} - \frac{\partial B_y}{\partial z} \\ &= \mu_0 J_x \end{aligned}$$

where we have applied a slightly manipulated version of Equation (6.4) in the final step. Similarly,

$$\begin{aligned} F^{2\nu}{}_{,\nu} &= \frac{\partial F^{20}}{\partial x^0} + \frac{\partial F^{21}}{\partial x^1} + \frac{\partial F^{22}}{\partial x^2} + \frac{\partial F^{23}}{\partial x^3} \\ &= -\frac{1}{c^2}\frac{\partial E_y}{\partial t} - \frac{\partial B_z}{\partial x} + 0 + \frac{\partial B_x}{\partial z} \\ &= \mu_0 J_y \end{aligned}$$

and

$$\begin{aligned} F^{3\nu}{}_{,\nu} &= \frac{\partial F^{30}}{\partial x^0} + \frac{\partial F^{31}}{\partial x^1} + \frac{\partial F^{32}}{\partial x^2} + \frac{\partial F^{33}}{\partial x^3} \\ &= -\frac{1}{c^2}\frac{\partial E_z}{\partial t} + \frac{\partial B_y}{\partial x} - \frac{\partial B_x}{\partial y} + 0 \\ &= \mu_0 J_z \end{aligned}$$

These are the results when we take the derivatives with respect to the second index on $F^{\mu\nu}$, differentiating across the rows. What if we differentiate across the columns with $F^{\mu\nu}{}_{,\mu}$? This is a simple matter of applying the antisymmetry: $F^{\mu\nu}{}_{,\mu} = -F^{\nu\mu}{}_{,\mu}$. Upon careful inspection, we find that

$$F^{\nu\mu}{}_{,\mu} = \mu_0 J^\nu = \mu_0 \begin{pmatrix} c\rho \\ J_x \\ J_y \\ J_z \end{pmatrix}$$

or, the derivative across one index of the electromagnetic field strength tensor provides the relativistic current density four-vector. The time component of this vector is directly related to the charge density producing the electric field. Tracking the subscripts while taking another derivative, we find that

$$F^{\mu\nu}{}_{,\nu,\mu} = \mu_0 J^\mu{}_{,\mu} = 0 \tag{6.5}$$

which is equivalent to Equation (6.1) and Equation (6.4) combined.

A similar equation exists for the dual tensor.

Chapter 7

The Postulate of General Relativity

7.1 Terminology: Special and General

What makes "special relativity" special? What makes "general relativity" general? These are common questions when one first researches the subjects, particularly when one is coming from a non-mathematical background.

In mathematical situations, research is often devoted to special and general cases. A "special" case is one in which specific values are known, or specific restraints are applied. For example, most elementary school math deals with special cases. Students solve questions such as 234×123 instead of solving the general case of $a \times b = c$ and then substituting $a = 234$ and $b = 123$ before solving for c.[1] Science works in a similar way. Special relativity considers the special case in which the reference frames one uses to measure from are experiencing uniform, or unaccelerated, motion. General relativity expands the theory to allow observers to move with arbitrary accelerations.

---

[1] It is the author's opinion that the lack of explicit instruction about the usefulness and need for general cases rather than exclusively special cases is one of the main reasons students struggle with algebra. When faced with a massive paradigm shift in mathematical thinking that is not *treated* as a massive paradigm shift for fear of scaring students away, the students become convinced that algebra is an exercise in futility and harder than it needs to be. When the author taught algebra to first time learners, he started with lessons about the paradigm shift and the axioms of algebra and the need for special cases before getting into the nitty-gritty steps. He then gave out a homework assignment asking students to manipulate high school physics formulae in the general case only, and the class average was 87%. When he later taught high school physics to a completely different group of students and gave those students the identical assignment, the class average was 63%. End rant.

7.2   The Postulate of General Relativity

The distinction between special and general relativity boils down to a single realization: one cannot distinguish between different accelerations by feel alone. Imagine being placed in a sealed and soundproof room (with ample oxygen supply.) Imagine also that you can distinguish between "up" and "down" as directions. There is a particular side of this room that you can call a "floor." Einstein realized[2] that there is no physical sensation that can be used to distinguish between standing in a sealed room on the surface of a planet and standing in a sealed room with a rocket engine attached that forces the room to accelerate.[3] This equivalence led to the formulation of non-inertial reference frames, or accelerated reference frames.

7.3   Gravity

Beyond allowing the formulation of non-inertial reference frames, this postulate connected gravity to reference frames, which were already connected to the speed of light. This was a monumental connection. At the time, only three forces were recognized in science: there was the electrical force, the magnetic force and the gravitational force. With the first two bound by relativity, scientists began searching for a unified theory that could bind the "final" force to the other two.[4] Developing this "Grand Unified Theory," or GUT, became Einstein's unrealized goal in life. Over 50 years after his death, such a theory has still not been formulated to the satisfaction of most physicists.

This formulation did, however, spawn a new revelation: mass and energy do not merely exist "in" the geometry of reality, they alter and define the geometry of reality. Imagine you are still in your earliest school years, and have been tasked with drawing a triangle. The implications of Einstein's postulate are akin to watching your piece of paper change and transform in shape and size as you draw the triangle on its surface. A flat piece of paper ceases to be flat as soon as something is drawn on it, instead curving in a way that attracts all other shapes on the page to the one you have just drawn. It is this revelation that has led to the notion of the "fabric" of spacetime. The mental image conjured by the mathematics renders reality as a great sheet. Massive and/or energetic objects dent this sheet, altering the "fabric" of the sheet of reality.

---

[2]He was riding an elevator at the time.

[3]We are forced to assume that any noises or vibrations caused by the rocket engine are not transmitted to the room.

[4]By today's count, there are four forces: the gravitational force, the electromagnetic force, the weak nuclear force and the strong nuclear force. Because electricity and magnetism are already connected and unified in our day to day situations, they are now treated as a single force. They have been unified with the weak and strong nuclear forces as well, but this unification is only visible under the extreme conditions created within particle accelerators, and so they are still treated as distinct forces in most circumstances.

7.4 The Geometry of Reality

This, once again, changed the way the geometry of the world was laid out. "Rotating" objects to different speeds caused ripples in the sheet, moving and shifting the dents in what we perceive as gravity. This even led to the theory of gravitational waves, which emanate from a source of gravity and propagate throughout the universe.

These dents and ripples have a great impact on the motion of objects. It is no longer natural to think of objects thinking in straight lines. They can move in the straightest possible lines, but that isn't quite the same thing; one can drive a "straight" stretch of road for several kilometers, but odds are the road curves along the surface of the Earth as elevations change on the oblate spheroid we call home. We now talk about "geodesics," which are the straightest possible lines one can follow on a curved surface.

To thoroughly drive the strangeness of this concept home, we will look at a shape that cannot exist on a flat surface, but which every reader is likely familiar with without even realizing that this is the case. The shape we are talking about is a "biangle." This shape is formed when two lines are drawn as straight as possible (i.e. two geodesics) and meet in exactly two places. In the geometry taught in public school, this cannot happen, but the geometry taught in public school is restricted to the "special case" of flat surfaces. The "general case" of geometry allows this shape to exist. You will probably not find examples in your mathematics classroom, but are extremely likely to find them in your geography classroom.

Take a globe of the Earth. It is covered with lines of longitude and latitude. Some, but not all, of these lines are geodesics. We need to determine which are which. Imagine you have a globe from the factory before it has been painted or marked in any way. It is a perfect sphere. You take a marker and start drawing the straightest possible line that you can from any point on the surface. You will eventually return to the spot you started from.[5] As the globe started with no distinguishing features, all such geodesic lines would be identical when drawn carefully and accurately. This is the case with lines of longitude on a globe, and not most[6] lines of latitude. If one takes two lines of longitude for examples on the globe, you will find that they form a biangle; they meet at the north pole, continue along the straightest possible paths until they are parallel while crossing the equator, and then converge to meet at the south pole and complete the biangle.

Similarly, "rules" for shapes that we are aware of do not necessarily apply. For example, we are taught that all triangles have angles which add up to 180°. While this is absolutely true on a flat surface, one can easily form a triangle with three right angles on an appropriately curved surface. Begin at any point on the equator. Travel directly to the north (or south) pole. Make a 90° turn and continue until you reach the equator once more. Make another 90° turn (in either direction, East

---

[5]Note that the author assumes your penmanship and artistic skills are significantly better than his own. He'd miss the target by a significant amount.

[6]There is a grand total of one line of latitude that forms a geodesic. Deducing which line that is indicates a good understanding of the topic.

or West) and return to your starting point. The triangle mapped out by these three geodesics will have three right angles, for a total angle of 270°.

We can take this same issue one step further and demonstrate the "parallel transport problem." Again, imagine you are standing on the equator, facing North with your right arm held out in the Eastward direction. Walk to the north pole, and with each step, you keep your arm held out parallel to the direction is was in when you took your previous step. Upon reaching the north pole, turn right, keeping your arm parallel to its original direction so that it now points directly ahead of you. Walk until you reach the equator and turn right again; you will now either have your right arm across the front of your body, or you will extend your left arm to maintain the same direction.[7] Walk back to your starting point. Your extended arm has been held parallel to its previous direction every step of the way, and yet the arm that started out pointing East is now pointing South! This is the parallel transport problem: you cannot transport something pointed a particular direction and guarantee that its direction will remain constant across a curved surface. Those paths which keep their own direction constant (i.e. those that don't "feel" like turning to a walking person: if you point your arm straight ahead with your first step, it points straight ahead of you at every step; if you return to your original position, you are facing in your original direction when you do) are the geodesics for that surface.

The concept of the geodesic is not just one that seems correct, but is one that turns out to be fundamental to the behaviour of objects in our universe. If we go back to Newton's original axioms, he stated "Unless an outside force is applied, a stationary object (object "at rest") will remain stationary and a moving object will continue moving at the same speed and in the same direction." In short, things move in straight lines. We now modify this to say that, in the absence of outside forces, objects move along geodesics.

This counter intuitive geometry had been explored by abstract mathematicians for centuries before its application to reality through relativity. With this toolkit available, physicists soon discovered theoretical objects that still hold imaginations enthralled to this day: black holes and worm holes.

7.5 Our New Coordinate Axes

The implications of Einstein's postulate of general relativity impact the geometry of space-time. Those implications are reflected in our mathematical constructs through the metric tensor. To this point, our metric has been defined as

$$g^{\mu\nu} = g_{\mu\nu} = \begin{pmatrix} -1 & 0 & 0 & 0 \\ 0 & 1 & 0 & 0 \\ 0 & 0 & 1 & 0 \\ 0 & 0 & 0 & 1 \end{pmatrix}$$

None of the entries in this metric are variable in any way. This implies that our Universe is completely uniform and unchanging, and that space is not affected by the arrangement of objects within it. We now realize this is not the case. Thus, we

---

[7]Your right arm is probably pretty sore at this point.

need to construct a new metric that represents real life situations. This brings us even further from Euclidean space.

To this point, our reference frames have been based on four axes: ct, x, y and z. These will no longer be the most natural axes. Space is distorted by gravity, and the most easily studied sources of gravity are stars, planets, moons, and similar astrophysical objects. These tend to share one thing in common: their self-gravitational force is powerful enough to ensure that they are reshaped into approximately spherical objects. Thus, the spherical coordinate system is the one that seems appropriate to most situations.[8]

In the spherical coordinate system, our imaginary axis ct will remain unchanged. We will divide the spatial coordinates into three new directions, which are most easily imagined using a globe of the Earth. Pick a point anywhere on the Earth's surface. The three coordinates are described by numbers r, θ and ϕ. The r coordinate is the radial axis, describing the distance from the centre of the Earth to the point chosen. In this case, the value of r would be the radius of the Earth $r_E = 6.38 \times 10^6 m$. The direction is along the line from the centre of the Earth to the selected point, which means the direction is variable. We've been dealing with coordinate systems that change all along, but to this point, they have only changed as they move with respect to each other, and not with respect to points measured by the coordinate system. This is no longer an option in general relativity; our coordinate systems must have the freedom to move around. It is customary to keep r as a non-negative number. It may be either zero or positive; negative values are represented by positive r values with different values for the angular coordinates we will now define.

The second coordinate, θ, is an angle describing the elevation relative to the line connecting the north and south poles. On a globe of the Earth, this is represented by the line of latitude involved. Specifically, it is measured relative to the line connecting the centre of the Earth with the north pole, and it is always taken to be a non-negative number. Thus, the tropic of Cancer (23.45° north latitude) would be at 66.55° if this angle were to be measured in degrees. The equator would be at 90°, and the tropic of Capricorn would be at 113.45°. The south pole would be at 180°, the largest possible value. The direction of this axis is parallel to Earth's surface, pointing south, the direction of increasing ϕ.

The third coordinate, ϕ, is an angle that measures rotation with respect to some arbitrary point around the surface, represented on Earth by lines of longitude. Angles are measured with respect to the Greenwich Mean Line, and rotate around the Earth from there. While Earth's geographers chose a standard that works with positive and negative values (marked "East" and "West"), mathematicians and physicists prefer to keep all values positive. If one looks down on the Earth from above the North Pole, then the coordinate ϕ increases as you move counter clockwise around the globe. The allowed values of ϕ would range from zero to 360°, representing a full circle, if they were measured in degrees at all; more on that later. The direction of this axis is parallel to the line of latitude on the surface of the Earth in the direction of increasing ϕ, which is East on the surface of the Earth.

---

[8]As we shall see, the best coordinate system is one created specifically for this application. It grows out of the spherical coordinate system, so that is the one we shall use.

The metric describing this system is one which must be derived. Before this is possible, we must work through a number of intermediate steps.

7.6   Radian Angular Measure

Angles are frequently measured in degrees in our day to day experiences, ranging from 0° to 360°. This is an ancient and convenient convention, chosen arbitrary by the humans who developed it.[9] This is also a system that has a deep but subtle flaw. When performing algebra with angles, the variable cannot appear outside a trigonometric function. In normal day to day applications, such as the whole of engineering, instances in which angles should appear outside trigonometric functions are extraordinarily rare, but in general relativity and theoretical mathematics, they occur regularly enough that we need a more natural means to measure angles. This natural angle is the radian.

The circle is a well known geometric shape. Every circle comes in the same proportions: the circumference around the outside C is directly proportional to the radius r through the relationship $C = 2\pi r$. This relationship forms the basis of the most natural angular measure. Imagine a circle with points a and b on the circumference:

---

[9]It is frequently taught that the 360 point limit was chosen to approximate the angle Earth swept through its orbit in one day. While this sounds reasonable, it does not appear to be the case; the earliest records of the 360 degree system are found in cultures that did not use the 365 day calendar. The cultures lacked fractions, so it appears 360 was chosen because a single degree is large enough to measure, but the circle can still be divided by a large number of divisors, as 360 is divisible by 1, 2, 3, 4, 5, 6, 8, 9, 10, 12, 15, 18 and 20.

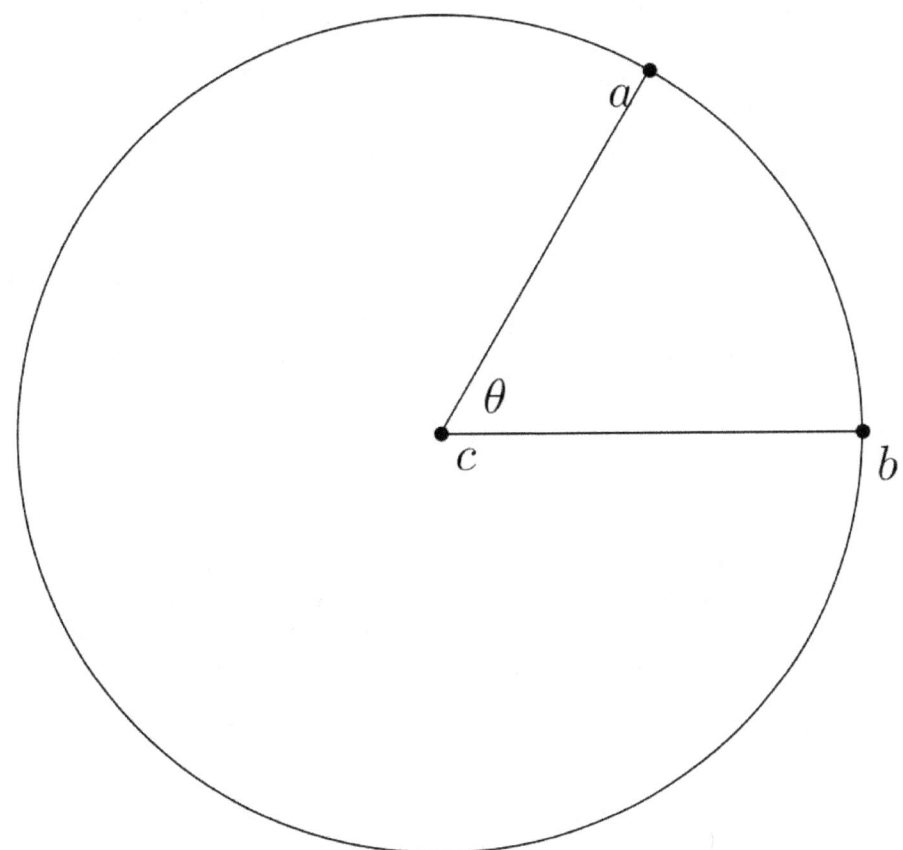

The length of the segment of the circumference s between points a and b is determined by two variables: the radius r, and the angle θ. Thus, radians are defined by the relationship between this segment's length and the radius of the circle:

$$\theta = \frac{s}{r}$$

Thus, the total angle in the centre of a circle is the angle that one gets when the segment is the full circumference of the circle:

$$\theta = \frac{2\pi r}{r} = 2\pi$$

Thus,

$$2\pi\text{rad} = 360°$$

or

$$l = \frac{360°}{2\pi\text{rad}} = \frac{180°}{\pi\text{rad}}$$

which is the conversion factor that is familiar to those who have seen radian measures before.

7.7   General Lorentz Transformation Tensors

The Lorentz transformation tensor we've seen previously

$$\Lambda^{\mu'}{}_{\nu} = \begin{pmatrix} \gamma & -\frac{v}{c}\gamma & 0 & 0 \\ -\frac{v}{c}\gamma & \gamma & 0 & 0 \\ 0 & 0 & 1 & 0 \\ 0 & 0 & 0 & 1 \end{pmatrix}$$

can be generalized into a tensor that converts between *any* two coordinate systems. We can convert from the system with basis vectors e_{ν} to the system with basis vectors $e_{\mu'}$ by defining each component of the Lorentz transformation tensor as

$$\Lambda^{\mu'}{}_{\nu} = \frac{\partial \mu'}{\partial \nu}$$

for each possible value of μ' and ν. To see how this is applicable, we will calculate the conversion factors from three dimensional Cartesian coordinates (ν) to three dimensional spherical coordinates (μ'). This requires first defining r, ϕ and θ in terms of x, y and z, and vice versa.

It is easiest to define the Cartesian coordinates in terms of the spherical coordinates:

$$x = r \sin \theta \cos \phi$$
$$y = r \sin \theta \sin \phi$$
$$z = r \cos \theta$$

We can manipulate these equations to find the spherical coordinates in relation to the Cartesian coordinates:

$$r = \sqrt{x^2 + y^2 + z^2}$$
$$\theta = \cos^{-1}\left(\frac{z}{\sqrt{x^2 + y^2 + z^2}}\right)$$
$$\phi = \tan^{-1}\left(\frac{y}{x}\right)$$

We can now define the general Lorentz transformation tensors that connect our two coordinate systems. We start with the tensor that converts Cartesian coordinates to spherical coordinates. As basis vectors have lowered indices, this would be the transformation

$$\Lambda^{\mu}{}_{\nu'}\mathbf{e}_{\mu} = \mathbf{e}_{\nu'}$$

where primed indices correspond to the spherical coordinate system. Thus,

$$\Lambda^{\mu}{}_{\nu'} = \begin{bmatrix} \Lambda^{x}{}_{r} & \Lambda^{x}{}_{\theta} & \Lambda^{x}{}_{\phi} \\ \Lambda^{y}{}_{r} & \Lambda^{y}{}_{\theta} & \Lambda^{y}{}_{\phi} \\ \Lambda^{z}{}_{r} & \Lambda^{z}{}_{\theta} & \Lambda^{z}{}_{\phi} \end{bmatrix}$$

$$= \begin{bmatrix} \frac{\partial x}{\partial r} & \frac{\partial x}{\partial \theta} & \frac{\partial x}{\partial \phi} \\ \frac{\partial y}{\partial r} & \frac{\partial y}{\partial \theta} & \frac{\partial y}{\partial \phi} \\ \frac{\partial z}{\partial r} & \frac{\partial z}{\partial \theta} & \frac{\partial z}{\partial \phi} \end{bmatrix}$$

$$= \begin{bmatrix} \sin\theta\cos\phi & r\cos\theta\cos\phi & -r\sin\theta\sin\phi \\ \sin\theta\sin\phi & r\cos\theta\sin\phi & r\sin\theta\cos\phi \\ \cos\theta & -r\sin\theta & 0 \end{bmatrix}$$

Thus,

$$\mathbf{e}_r = \Lambda^{\mu}{}_r \mathbf{e}_{\mu}$$
$$= \frac{\partial x}{\partial r}\mathbf{e}_x + \frac{\partial y}{\partial r}\mathbf{e}_y + \frac{\partial z}{\partial r}\mathbf{e}_z$$
$$= \sin\theta\cos\phi\,\mathbf{e}_x + \sin\theta\sin\phi\,\mathbf{e}_y + \cos\theta\,\mathbf{e}_z$$

Similarly,

$$\mathbf{e}_{\theta} = r\cos\theta\cos\phi\,\mathbf{e}_x + r\cos\theta\sin\phi\,\mathbf{e}_y - r\sin\theta\,\mathbf{e}_z$$
$$\mathbf{e}_{\phi} = -r\sin\theta\sin\phi\,\mathbf{e}_x + r\sin\theta\cos\phi\,\mathbf{e}_y$$

We can construct the metric of spherical coordinates knowing that

$$g_{\mu\nu} = \mathbf{e}_{\mu} \cdot \mathbf{e}_{\nu}$$

and find

$$g_{\mu\nu} = \begin{bmatrix} \mathbf{e}_r \cdot \mathbf{e}_r & \mathbf{e}_r \cdot \mathbf{e}_{\theta} & \mathbf{e}_r \cdot \mathbf{e}_{\phi} \\ \mathbf{e}_{\theta} \cdot \mathbf{e}_r & \mathbf{e}_{\theta} \cdot \mathbf{e}_{\theta} & \mathbf{e}_{\theta} \cdot \mathbf{e}_{\phi} \\ \mathbf{e}_{\phi} \cdot \mathbf{e}_r & \mathbf{e}_{\phi} \cdot \mathbf{e}_{\theta} & \mathbf{e}_{\phi} \cdot \mathbf{e}_{\phi} \end{bmatrix}$$

$$= \begin{bmatrix} 1 & 0 & 0 \\ 0 & r^2 & 0 \\ 0 & 0 & r^2\sin^2\theta \end{bmatrix}$$

The reverse transformation is facilitated by

$$\Lambda^{\mu'}{}_{\nu} = \begin{bmatrix} \Lambda^{r}{}_{x} & \Lambda^{r}{}_{y} & \Lambda^{r}{}_{z} \\ \Lambda^{\theta}{}_{x} & \Lambda^{\theta}{}_{y} & \Lambda^{\theta}{}_{z} \\ \Lambda^{\phi}{}_{x} & \Lambda^{\phi}{}_{y} & \Lambda^{\phi}{}_{z} \end{bmatrix}$$

$$= \begin{bmatrix} \frac{\partial r}{\partial x} & \frac{\partial r}{\partial y} & \frac{\partial r}{\partial z} \\ \frac{\partial \theta}{\partial x} & \frac{\partial \theta}{\partial y} & \frac{\partial \theta}{\partial z} \\ \frac{\partial \phi}{\partial x} & \frac{\partial \phi}{\partial y} & \frac{\partial phi}{\partial z} \end{bmatrix}$$

For the sake of fitting the results onto the page, each column of the result will be listed individually.

$$\Lambda^{\mu'}{}_x = \begin{bmatrix} \dfrac{x}{\sqrt{x^2+y^2+z^2}} \\[2ex] \dfrac{xz}{(x^2+y^2+z^2)^{\frac{3}{2}} \sin\left(\cos^{-1}\left(\frac{z}{\sqrt{x^2+y^2+z^2}}\right)\right)} \\[4ex] -\dfrac{y}{x^2 \sec^2\left(tan^{-1}\left(\frac{y}{x}\right)\right)} \end{bmatrix}$$

$$\Lambda^{\mu'}{}_y = \begin{bmatrix} \dfrac{y}{\sqrt{x^2+y^2+z^2}} \\[2ex] \dfrac{yz}{(x^2+y^2+z^2)^{\frac{3}{2}} \sin\left(\cos^{-1}\left(\frac{z}{\sqrt{x^2+y^2+z^2}}\right)\right)} \\[4ex] \dfrac{1}{x^2 \sec^2\left(tan^{-1}\left(\frac{y}{x}\right)\right)} \end{bmatrix}$$

$$\Lambda^{\mu'}{}_z = \begin{bmatrix} \dfrac{z}{\sqrt{x^2+y^2+z^2}} \\[2ex] \dfrac{z\left(x-\sqrt{x^2+y^2+z^2}\right)}{(x^2+y^2+z^2)^{\frac{3}{2}} \sin\left(\cos^{-1}\left(\frac{z}{\sqrt{x^2+y^2+z^2}}\right)\right)} \\[4ex] 0 \end{bmatrix}$$

which is far simpler to calculate than it would appear. We are now equipped to convert between the two coordinate systems. This is not sufficient for work in relativity, though. We often wish to work in two coordinate systems simultaneously, such as those in our S and S' frames. This is the framework that we still need to establish.

7.8   Christoffel Symbols

Let us examine

$$\frac{\partial \mathbf{e}_x}{\partial \theta}$$

in detail. As $\mathbf{e}_x$ is constant, the derivative should come to zero regardless of coordinate system.[10] Using our above expressions written in spherical coordinates, we find that

$$\mathbf{e}_x = \frac{\partial r}{\partial x}\mathbf{e}_r + \frac{\partial \theta}{\partial x}\mathbf{e}_\theta + \frac{\partial \phi}{\partial x}\mathbf{e}_p hi$$

$$= \frac{x}{\sqrt{x^2+y^2+x^2}}\mathbf{e}_r + \frac{xz}{(x^2+y^2+z^2)^{\frac{3}{2}} \sin\left(\cos^{-1}\left(\frac{z}{\sqrt{x^2+y^2+z^2}}\right)\right)}\mathbf{e}_\theta$$

$$- \frac{y}{x^2 \sec^2\left(tan^{-1}\left(\frac{y}{x}\right)\right)}\mathbf{e}_\phi$$

$$= \sin\theta\cos\phi\mathbf{e}_r + \frac{\cos\theta\cos\phi}{r}\mathbf{e}_\theta - \frac{\sin\phi}{r\sin\theta}\mathbf{e}_\phi$$

---

[10]Remember, the spherical and Cartesian coordinates do not move with respect to one another. The spherical unit vectors are not constant, but they don't move *relative to one another*.

One might then assume

$$\frac{\partial \mathbf{e}_x}{\partial \theta} = \cos\theta \cos\phi \mathbf{e}_r - \frac{\sin\theta \cos\phi}{r}\mathbf{e}_\theta \frac{\cos\theta \sin\phi}{r \sin^2\theta}\mathbf{e}_\phi$$

but one would be wrong. The problem is that the basis vectors are not constant. Thus, the proper derivative is

$$\frac{\partial \mathbf{e}_x}{\partial \theta} = \cos\theta \cos\phi \mathbf{e}_r - \frac{\sin\theta \cos\phi}{r}\mathbf{e}_\theta \frac{\cos\theta \sin\phi}{r \sin^2\theta}\mathbf{e}_\phi$$
$$+ \sin\theta \cos\phi \frac{\partial \mathbf{e}_r}{\partial \theta} + \frac{\cos\theta \cos\phi}{r}\frac{\partial \mathbf{e}_t heta}{\partial \theta} - \frac{\sin\phi}{r \sin\theta}\frac{\partial \mathbf{e}_\phi}{\partial \theta}$$

It is this complete combination that is identically zero, as

$$\frac{\partial \mathbf{e}_r}{\partial \theta} = \frac{1}{r}\mathbf{e}_\theta$$
$$\frac{\partial \mathbf{e}_\theta}{\partial \theta} = -r\mathbf{e}_r$$
$$\frac{\partial \mathbf{e}_\phi}{\partial \theta} = \frac{\cos\theta}{\sin\theta}\mathbf{e}_\phi$$

The final details of the proof left to the reader. It is a matter of a large number of trigonometric gymnastics.

The most important point to take away from the above is that systems with variable basis vectors have more involved vector and tensor derivatives than other systems. Thus, we introduce the Christoffel symbols to facilitate such derivatives:

$$\Gamma^\mu{}_{\alpha\beta}\vec{e}_\mu = \frac{\partial \vec{e}_\alpha}{\partial x^\beta} \tag{7.1}$$

Just as we defined

$$A^\mu{}_{,\nu} = \frac{\partial A^\mu}{\partial x^\nu}$$

in Cartesian space with its constant basis vectors, we can now define

$$A^\mu{}_{;\nu} = A^\mu{}_{,\nu} + A^\alpha \Gamma^\mu{}_{\alpha\beta}$$

as the most useful total derivative in systems with variable basis vectors.

It should be noted that this definition of Christoffel symbols can often be extremely cumbersome. At times, it is easier to computer the Christoffel symbols using the identity

$$\Gamma^\gamma{}_{\beta\mu} = \frac{1}{2}g^{\alpha\gamma}\left(g_{\alpha\beta,\mu} + g_{\alpha\mu,\beta} - g_{\beta\mu,\alpha}\right) \tag{7.2}$$

Similarly, the general divergence of a vector can be defined as

$$\vec{\Box} \cdot \vec{A} = \Box_\mu A^\mu = A^\mu{}_{;\mu} = A^\mu{}_{,\mu} + \Gamma^\alpha{}_{\mu\alpha}A^\mu$$

7.9   Pseudo-Riemannian Manifolds

We can now start to examine manifolds. A "manifold" in mathematics is a surface within a higher dimensional space. If you take a sphere of a given radius, then the surface of that sphere is a two dimensional manifold: the r coordinate can be assumed, so the position on the surface can be described completely by the two coordinates θ and ϕ. A sufficiently small observer standing on the surface (such as humans on the surface of the Earth) could easily mistake this surface for a flat one. This last property, in which "local" coordinates appear to fit Minkowski metrics, make the surface pseudo-Riemannian. A Riemannian manifold is one in which every vector dot product (i.e. $A^\mu A_\mu$) is positive definite for non-zero vectors; this property never applies in relativity, as light-like vectors have 0 dot products and time-like vectors have negative dot products (in our chosen convention.) The full derivation and definition of such objects will be quite abbreviated in this treatment, omitting most proofs and derivations for the sake of brevity.[11] The basic mechanism of the proof is similar to working out the first derivative in calculus from first principles; set up a surface, look at points (x, y), $(x + \delta x, y)$, $(x, y + \delta y)$ and $(x + \delta x, y + \delta y)$, and then work out the vector differences between the two points and the surface vectors that describe them.

It can be shown that a vector V^α is kept parallel to its original direction as it is carried along a curve U^β if the equation

$$U^\beta V^\alpha{}_{;\beta} = 0$$

holds along the entire curve. The curve U^β is a geodesic if it can transport its tangent vector in this fashion, such that

$$U^\beta U^\alpha{}_{;\beta} = U^\beta U^\alpha{}_{,\beta} + \Gamma^\alpha{}_{\mu\beta} U^\mu U^\beta = 0 \qquad (7.3)$$

This places constraints on U^β that depend upon the Christoffel symbols. The Christoffel symbols, in turn, depend on the metric, and the metric defines the curvature of the space-time or manifold.

Sufficient algebraic gymnastics can be performed to define the curvature of the space in terms of the *Riemannian tensor*. This tensor is most naturally defined in terms of the Christoffel symbols, but can also be defined in terms of the metric, as follows:[12]

$$R^\alpha{}_{\beta\mu\nu} = \Gamma^\alpha{}_{\beta\nu,\mu} - \Gamma^\alpha{}_{\beta\mu,\nu} + \Gamma^\alpha{}_{\sigma\mu}\Gamma^\sigma{}_{\beta\nu} - \Gamma^\alpha{}_{\sigma\nu}\Gamma^\sigma{}_{\beta\mu}$$

This Riemannian tensor is constructed of combinations of the symmetric metric tensor which differ by signs. That means it has a number of symmetries and antisymmetries of its own. These can be summarized with a couple of identities, once one has established that $R_{\alpha\beta\mu\nu} = g_{\alpha\sigma}R^\sigma{}_{\beta\mu\nu}$ is a valid way to lower the index on

---

[11]Readers looking for more complete details are encouraged to track down *A First Course in General Relativity* by Bernard F. Schutz, ISBN 0-521-27703-5.[3]

[12]If you ever meet someone who can write and recognize the entire Greek alphabet but has no idea what sounds the letters make or what order they come in, that person has probably studied physics.

$R^\alpha{}_{\beta\mu\nu}$:

$$R_{\alpha\beta\mu\nu} = -R_{\beta\alpha\nu\mu} = -R_{\alpha\beta\nu\mu} = R_{\mu\nu\alpha\beta}$$
$$R_{\alpha\beta\mu\nu} + R_{\alpha\nu\beta\mu} + R_{\alpha\mu\nu\beta} = 0$$

The specific entries in this tensor are the ones that describe the curvature in space. In space that isn't curved, the Christoffel symbols are all 0, and so is the Riemann tensor. Every straight line is a geodesic and parallel lines are preserved.

7.10 The Metric of Newtonian Gravity

We are finally equipped to create the equations that describe curvature in space that has been generated by Newtonian gravity.[13] We know that Newton's gravitational equations work well for low speeds and relatively weak gravitational sources, as they have worked very well for years. We also know, thanks to Mercury's orbit, that they are not a complete description. In our first step, we construct a metric which gives us Newtonian gravity to first order, and then we will generalize to other cases later. One of the assumptions that we are required to make when constructing the metric is that the gravitational field does not change with time. This will turn out to be the vital difference between Newtonian and relativistic gravity.

Generally speaking, gravity is a weak force. We construct our approximation theory of gravity by adding terms related to the strength of the field f to the Minkowski metric as follows:

$$g_{\mu\nu} = \begin{pmatrix} -1 - O(f) - O(f^2) - \cdots & 0 & 0 & 0 \\ 0 & 1 + O(f) + O(f^2) + \cdots & 0 & 0 \\ 0 & 0 & 1 & 0 \\ 0 & 0 & 0 & 1 \end{pmatrix}$$

where $O(f^n)$ is read "terms of order f^n" and means "terms that include a factor of f^n". For Newtonian gravity, we will only concern ourselves with terms of order f, and in which all higher powers are negligible in our day to day lives.[14] The field strength should remain unchanged when compared to its Newtonian form, so

$$f = -\frac{GM}{r}$$

where $G = 6.67 times 10^{-11} \frac{Nm^2}{kg^2}$ is Newton's gravitational constant, M is the total inertia of the object rather than its mass, and r is the distance from that source of mass/inertia M. If we are adding this term to the dimensionless constant 1, or subtracting it from -1, we need to apply a coefficient to render this dimensionless. The overall units on f are

$$\frac{Nm}{kg} = \frac{kgm^2}{s^2kg} = \frac{m^2}{s^2}$$

---

[13] It truly will be more "creation" than derivation. A full derivation would require either more than nine chapters in this book, or absolutely gargantuan chapters.

[14] This is not an unjustified assumption: this is exactly what we already have when comparing Newtonian equations for kinetic energy, velocity, acceleration, etc. to their relativistic counterparts.

which is a strong indication of the constant to use. With the addition of a factor of 2, justified solely by testing this result and finding that the factor is required to conform to Newtonian gravitational results, we arrive at the final form of our metric:

$$g_{m u v} = \begin{pmatrix} -1 + \frac{2GM}{c^2 r} & 0 & 0 & 0 \\ 0 & 1 - \frac{2GM}{c^2 r} & 0 & 0 \\ 0 & 0 & 1 & 0 \\ 0 & 0 & 0 & 1 \end{pmatrix}$$

If we place our source of gravity at the origin of our coordinate system, then the metric simplifies to

$$g_{\mu v} = \begin{pmatrix} -1 + \frac{2GM}{c^2 \sqrt{x^2+y^2+z^2}} & 0 & 0 & 0 \\ 0 & 1 - \frac{2GM}{c^2 \sqrt{x^2+y^2+z^2}} & 0 & 0 \\ 0 & 0 & 1 & 0 \\ 0 & 0 & 0 & 1 \end{pmatrix} \qquad (7.4)$$

To see how this describes the action of bodies, we recall our modified axiom: objects which experience no outside forces move along geodesic curves, conforming to the geodesic Equation (7.3). Imagine a body at rest with position vector

$$\vec{x} = \begin{pmatrix} 0 \\ x \\ 0 \\ 0 \end{pmatrix}$$

We now track the behaviour of this object over time through the geodesic equation. We force $x^\alpha{}_{,\beta} = 0$ to be true, which leaves the constraint

$$x^\alpha{}_{,\beta} + \Gamma^\alpha{}_{\mu\beta} x^\mu x^\beta = 0$$

or (given that x^1 is the only component of x^μ which is not zero)

$$\frac{d}{d\tau} x = -\Gamma^1{}_{11} x^1 x^1$$

By Equation (7.2),

$$\Gamma^1{}_{11} = \frac{1}{2} g^{11} \left(g_{11,1} + g_{11,1} - g_{11,1} \right) = \frac{1}{2} g^{11} g_{11,1}$$

which results in the expression

$$\frac{dx}{d\tau} = -\frac{GMc}{c^2 r - 2GM}$$

which implies exactly what we wanted: an object sitting freely in the vacuum will move to the origin of the coordinate system whether a force acts upon it or not. The shape of the space itself demands movement toward the origin. Gravity is now built into the shape of the universe at its most fundamental level.

This is probably the last time we will be using Cartesian coordinates as our main coordinate system. As soon as the gravitational field changes over time, some of our assumptions break down. Furthermore those broken assumptions mean we can no longer ignore the higher order terms (which, for terms of order f^n, includes terms of order c^{-2n}, which is why we don't notice them) as speeds really start to approach those of light, and our dimensionless higher order terms start to take on values that are closer and closer to 1. These effects will be examined closely in the context of black holes and worm holes.

Chapter 8

Holes in Space

8.1 Black Holes

Scientists and science fiction writers have been fascinated by black holes since before they even had names. Some of these writers did insufficient research, and as a result, popular understanding of black holes includes a number of misconceptions. Before we begin to examine the reality and the fiction surrounding these objects, however, we must first establish some quantum mechanics.

8.1.1 Quantum Mechanics

All objects with mass are made up of a collection of subatomic particles, the most common of which are electrons, protons and neutrons. It is less common knowledge that protons and neutrons are, themselves, made up of subatomic particles known as quarks, and that the right conditions can force a proton and electron to combine to form a neutron. The universe is also filled with "virtual particles," which appear temporarily, generally in pairs, and allow objects to feel forces, among other things.

As surprising as it may be, these incredibly tiny objects serve important roles in the study of black holes. We will deal with virtual particles in subsection 8.1.4. For now, we will only examine the effects of great pressure on atoms and nuclei.

Stars are the most massive common objects in our galaxies. They release energy due to natural fusion reactions. In short, they have so much mass that the atoms in their cores are crushed together into larger particles. The heat energy this releases helps the star maintain its size and structure. As the atoms within the star get larger and larger, it requires more and more energy to force them to combine into larger atoms through fusion. Stars eventually "run out of fuel," meaning that the atoms in their core are too large to be forced to fuse together by the star's gravity any longer. The star "dies" as the internal pressure suddenly drops, setting off a chain reaction that leads to a nova or supernova explosion as the internal pressure suddenly drops. What remains is a massive husk, rife with gravitational pressures, but lacking the internal fusion that provides a counter pressure to keep the star at its original size. If this husk is large enough, it forms a neutron star; all of its internal atoms are gravitationally crushed together until the core is a single giant nucleus made solely out of neutrons. In some cases, the husk has too much mass to even exist as a neutron star, and it continues to collapse. These objects become black holes.

8.1.2   What Black Holes Are and Aren't

A black hole is an object that has so much mass its internal gravity is too great for even a giant nucleus to hold itself together. All of the particles within it continue to collapse into a point, also known as a singularity. This singularity has mass, but no volume; it is just a spot in space. It is this lack of volume that makes it dangerous.

The force of gravity felt between two objects depends on three variables: the mass of one object, the mass of the other object, and the distance between them. The closer the two objects get, the stronger the force of gravity between them. This is why gravity is always strongest when the objects are in contact, and the distance between the centres of the two objects is as small as it can get. In the special case in which at least one object is a planet, the force of gravity is strongest when the second object is on the planet's surface. Imagine Earth is the first object and you are the second object. If you were to start digging into the Earth, travelling down into and through the molten mantle, you would notice that gravity steadily decreased as you got closer to the core.[1] This is because much of the Earth's mass is "above" you as you tunnel, and the gravitational force from that mass is pulling you upwards. Now imagine the Earth has collapsed into a black hole. It has no surface that you can reach, so getting closer and closer just means the force of gravity gets stronger and stronger.

The stronger the pull of gravity, the harder it is to escape. The faster you are travelling, the better your chance at escape. (Rockets can escape Earth's gravity to visit the moon and other places, but people can't escape Earth's gravity by running up a hill and jumping as high as they can.) The key to leaving a planet is reaching "escape velocity," the speed at which one must travel away from the planet's core in order to escape its gravity. A black hole is an object with no volume, which means there is no limit to how close one can get to it. As a result, one can get close enough to it that the escape velocity at that point is greater than the speed of light. The point at which escape velocity equals the speed of light is known as the "event horizon."[2] As we know, the speed of light is woven into the fabric of our spacetime. The geometry of black holes gives them their name: if the "dent" made by the gravity of an object is so steep that not even light can escape, then for all intents and purposes, it punches a "hole" in the fabric of reality. In actuality, the fabric of spacetime still exists within the black hole, but it might as well vanish as far as those outside the black hole are concerned, as the interior of the black hole can never be observed.

---

[1]This is assuming you notice anything other than the intolerable heat.

[2]The etymology of this name goes back to the mind set from earlier chapters: relativity doesn't focus on objects at locations in space, but at events that occur at locations in spacetime. Just as one cannot see objects past the horizon on Earth, one cannot observe events that take place behind the event horizon of a black hole.

It is a common belief that, should our Sun be replaced by a black hole with mass equal to our Sun, our entire solar system would be swiftly sucked into this black hole and crushed beyond recognition. This could not be further from the truth. The key to the unusual effects of a black hole is proximity; if you aren't close to the event horizon, you don't notice a difference. If our Sun were to be replaced with such a black hole in some instantaneous manner, Earth's orbit would be completely undisturbed. The lack of incoming energy from fusion would still doom the human race when photosynthesis stopped and temperatures dropped, but it would be a slow and agonizing planetary death instead of the swift and exciting death predicted by underresearched stories. When you do approach the event horizon closely enough, you witness a number of unusual phenomena.

8.1.3   Black Holes Have No Hair

There are a number of quantities that can be tracked and conserved when studying most objects and events. For example, if car A contains a certain number of electrons, and car B has a different number of electrons, the total number of electrons remains the same after the two cars collide in the street. One can count electrons after the fact and figure out how many electrons each vehicle started with.[3] With a black hole, that and other information is lost. In fact, when studying a black hole, only four pieces of information about the matter that forms the black hole are preserved: the total mass-energy of the objects and particles involved, the net electrical charge of those objects, the net magnetic charge[4] of the objects, and the net angular momentum[5] of those objects. It is said that "black holes have no hair," using the metaphor that the lost information is equivalent to hair. It's a strained metaphor.[6]

8.1.4   Hawking Radiation

One of the reasons Stephen Hawking is so highly regarded amongst the physics community is that he was one of the first to examine the combined implications of quantum mechanics and general relativity, and in doing so he made an astonishing discovery: black holes evaporate.

It is a well known fact of quantum mechanics that "virtual particles" appear and disappear frequently. Hawking was the first to examine what happens when these particles appear and disappear near a black hole. It is possible that one particle

---

[3]I don't claim it's easy to count them, only that it is conceptually possible.

[4]Although theoretically possible at quantum mechanical levels, magnetic monopoles (north poles without south poles or vice versa) have never been observed. In theory, if they do exist and fall into black holes, these quantities will survive.

[5]Angular momentum is a combination of an object's geometry and rotation speed. A black hole retains this combination, but isolated information about either the geometry or rotation speeds of the objects the black hole is made of is lost.

[6]If the metaphor is so strained, why is it in use? Purportedly, the scientists who proposed the vocabulary did so because several non-English languages translate the term "black hole" into a term that is also used as a slang term that refers to a part of the female anatomy. They thought it was amusing to force that phrasing upon scientists working in those languages. This kind of attitude may be one of the reasons the field of physics is still so male-dominated today. The gender gap is smaller than it used to be, but it is still far too wide.

falls into the black hole, while the other escapes. These "virtual particles" typically annihilate with each other and disappear to conserve the total energy available. When one is captured by a black hole, it gives the black hole a *negative* amount of energy. Thus, the other particle can continue to exist as a real particle, and the black hole *loses* energy. The smaller the black hole, the faster this takes place. This is what has some people in a panic about particle accelerators: they fear that the creation of tiny black holes will somehow threaten the planet, either because of the energy released by evaporating black holes, or because the tiny black holes themselves are going to somehow consume the planet.

We have already discussed the latter fear; black holes don't suddenly suck everything up. This is a very good thing, as all elementary particles (electrons, the quarks protons and neutrons are made of, etc.) have zero volume, meaning every subatomic particle in existence is a black hole. Not big ones, but black holes nonetheless. As for the former fear, energy is still conserved, so a black hole that is made out of six high energy particles can only release the energy contained in those six particles while evaporating. Thus, it can only release energy in quantities less than or equal to the energy that was already contained by the particle accelerator and detector before the black hole was created.

8.1.5 Frame Dragging, Event Horizons and Ergoregions

Mass and energy have been tied tightly to the very fabric of spacetime. We know that a moving object creates ripples of gravity in this spacetime. What happens if it rotates rapidly, creating a lot of motion-ripples (gravitational radiation) in a single location?

The answer is a phenomenon known as "frame dragging," and it is not limited to black holes. When a massive object rotates rapidly, it can pull spacetime along with it. If you are a stationary observer positioned above this source of gravity (likely a star) you would not fall straight toward the object. Instead, you would be "dragged" around the object in the same direction it is rotating as you fall, which is why the phenomenon is called frame dragging. This is the phenomenon that explains the unusual nature of Mercury's orbit, first mentioned way back on the first page of our first lesson. Mercury's orbit rotates around the Sun because the spacetime around the Sun rotates *with* the Sun, dragging our solar system's closest planet right along with it. This effect is most pronounced near the object that is rotating.

Perhaps the most bizarre phenomenon surrounding black holes takes place inside the event horizon. When an observer falls inside the black hole, spacetime gets so twisted that the four directions of space and time get twisted around. Once past the event horizon, an observer is just as free to move through the time dimension as we typically can move through space. The tradeoff is that the "down" direction takes on the position of the imaginary axis, meaning the observer is completely unable to control his or her inevitable descent towards the singularity at the centre of the black hole.

Because the effect of frame dragging is more pronounced as you get closer to the object, it becomes a significant effect when a black hole rotates. In fact, many black holes can rotate so quickly that they form "ergoregions" around them. In these regions, spacetime is being pulled so strongly that it becomes impossible to rotate around the black hole in the direction opposite to its rotation. If the black hole rotates clockwise (from your perspective) then it is impossible to form a close counterclockwise orbit around that black hole. Furthermore, the transformation and replacement of the "down" and "time" directions takes place within the ergoregion instead of behind the event horizon. When one falls through the ergoregion of one of these rapidly rotating black holes, and crosses the event horizon within, the transformation is reversed: within this black hole, time and space resume their normal roles! In fact, it is possible that the entire observable universe is contained within a supermassive, rapidly rotating black hole.

8.2 Wormholes

With the discovery that mass can warp spacetime, punching holes in it, scientists and science fiction writers got excited again. Perhaps it was possible to circumvent the light speed limit in some way: instead of propelling an object faster than light, one could bend and twist space until the destination is much closer, and then just move there at speeds less than light.

The first scientific suggestion related to this idea is commonly referred to as the "wormhole." Formally named the Einstein-Rosen bridge (as Albert Einstein and Nathan Rosen first proposed it in 1935 as a way to bridge two points in spacetime) the idea is that a tunnel, or hole, is created in spacetime by two black holes. Each black hole pulls the fabric of spacetime out of shape into an escapable region. It was proposed, however, that two such black holes, properly aligned, could create a tunnel that connected two far distant points in spacetime. Unfortunately, 27 years later, John Wheeler and Robert Fuller showed that such a body is unstable, and would collapse as soon as any mass or energy entered, turning the wormhole back into a pair of black holes before it could come to any practical use.

In 1988, Kip Thorne and Mike Morris showed that stable wormholes are only possible if some sort of exotic matter is held in place at the opening. While this does give some hope for faster than light travel, it is not much hope. The type of matter proposed has properties that have never been observed in our universe, and there have been no effective methods proposed for keeping that matter in place. Try as we might, it appears that the universe simply does not allow time travel to happen in the science fiction sense.

8.3 Schwarzschild Geometry

At the end of the last lesson, we assembled (but did not derive) a metric that approximated Newtonian behaviours and demonstrated that such a metric necessitated that objects fall under gravity. It was Karl Schwarzschild who, in 1915, found solutions to Einstein's equations that could be formally derived and possessed these properties. The metric he discovered is

$$
g_{\mu\nu} = \begin{pmatrix}
-\left(1 - \frac{2GM}{c^2 r}\right) & 0 & 0 & 0 \\
0 & \frac{1}{1 - \frac{2GM}{c^2 r}} & 0 & 0 \\
0 & 0 & r^2 & 0 \\
0 & 0 & 0 & r^2 \sin^2\theta
\end{pmatrix}
$$

Examination of this metric led to the discovery of black hole theory. In the case in which

$$
r = \frac{2GM}{c^2}
$$

the metric reduces to

$$
g_{\mu\nu} = \begin{pmatrix}
0 & 0 & 0 & 0 \\
0 & \infty & 0 & 0 \\
0 & 0 & r^2 & 0 \\
0 & 0 & 0 & r^2 \sin^2\theta
\end{pmatrix}
$$

which is something of an issue for the g_{00} and g_{11} entries. Is this a singularity of the geometry, or simply of the coordinate system? After all, the traditional spherical coordinate system has singularities at $\theta = 0°$ and $\theta = 180°$. Going back to our model of the Earth, if one is at either the north or south poles, you have a singularity of the coordinate system as there is no unique line of longitude (ϕ) to describe that location. Nonetheless, the location is perfectly valid, and the problem is with the coordinates, and not the space itself. An appropriate set of coordinates was discovered in 1960, known as Kruskal-Szekeres coordinates, and it eliminates the apparent issue here with the definitions

$$
u = e^{\frac{c^2 r}{4GM}} \cosh\left(\frac{c^3 t}{4GM}\right) \sqrt{\frac{c^2 r}{2GM} - 1}
$$

$$
v = e^{\frac{c^2 r}{4GM}} \sinh\left(\frac{c^3 t}{4GM}\right) \sqrt{\frac{c^2 r}{2GM} - 1}
$$

for $r > \frac{2GM}{c^2}$ and

$$
u = e^{\frac{c^2 r}{4GM}} \sinh\left(1 - \frac{c^3 t}{4GM}\right) \sqrt{\frac{c^2 r}{2GM}}
$$

$$
v = e^{\frac{c^2 r}{4GM}} \cosh\left(1 - \frac{c^3 t}{4GM}\right) \sqrt{\frac{c^2 r}{2GM}}
$$

for $r < \frac{2GM}{c^2}$. These definitions share the metric

$$
g_{\mu\nu} = \begin{pmatrix}
-\frac{32G^3 M^3}{c^6 r} e^{-\frac{c^2 r}{2GM}} & 0 & 0 & 0 \\
0 & \frac{32G^3 M^3}{c^6 r} e^{-\frac{c^2 r}{2GM}} & 0 & 0 \\
0 & 0 & r^2 & 0 \\
0 & 0 & 0 & r^2 \sin^2\theta
\end{pmatrix}
$$

where v is the 0th coordinate. Note that u and v are both unitless in this definition: the required units are "carried" by the metric components, which now all have units of length squared. Many find this aesthetically pleasing, as all components of the metric now have consistent units, and all components of all four vectors would also now have consistent units.

These coordinates handle the coordinate singularity, but they have some behaviour consistent with the Schwarzschild metric which is surprising. What happens when $r < \frac{2GM}{c^2}$? In both the Schwarzschild and Kruskal-Szekeres metrics, something strange occurs: $g_{00} > 0$ and $g_{11} < 0$ under these conditions. In other words, the time direction we are used to becomes as easy to navigate as space, but the *radial* direction that carries an object towards $r = 0$ marches inevitably forward! Once an observer gets too close to the point $r = 0$, it becomes an inescapable trap. This is a black hole, and the surface described by

$$ R_S = \frac{2GM}{c^2} $$

is known as the event horizon. The distance R_S is known as the Schwarzschild radius. It should be noted that the event horizon is only spherical in the case where the mass M is not rotating.

How dangerous is this? Well, for a body like our Sun, with $M = 1.98892 \times 10^{30} kg$, we have $R_S = 2953.25m$. This is over 235000 times smaller than the Sun's current radius, so we need not worry about our Sun becoming a black hole unless it is somehow compressed to under $\frac{1}{1.306 times 10^{16}}$ times its current volume. Notice also that the gravity of the black hole behaves just like the gravity of a normal object for $r \gg R_S$, which is the behaviour we would see in Earth's orbit if our Sun were replaced by an equally massive black hole.

8.3.1 What is at $r = 0$?

We have seen that the radial dimension takes on a timelike nature within the Schwarzschild radius. Thus, one cannot resist an immediate forward crush through this dimension. As a result, any mass within the Schwarzschild radius travels directly to that point in the spacetime. Furthermore, attempting to describe this point in terms of either metric results in division by zero. This is the *singularity* at the centre of a black hole. All mass within the black hole finds its way here. There is no way to distinguish between the different contributions to this singularity: hence the thought that "black holes have no hair." The only quantities which survive the crush are the total mass-energy, net electrical charge, net magnetic charge and total angular momentum. Thus far, we have ignored angular momentum, as it is not a part of the Schwarzschild geometry.

8.4 Kerr Geometry

The Kerr metric is one in which angular momentum is not necessarily zero, and the black hole is permitted to rotate. If the black hole carries angular momentum J, then the Kerr metric is given by

$$g_{\mu\nu} = \begin{pmatrix} -\frac{\Delta - a^2 \sin^2\theta}{\rho^2} & 0 & 0 & -\frac{2Mra\sin^2\theta}{\rho^2} \\ 0 & \frac{\rho^2}{\Delta} & 0 & 0 \\ 0 & 0 & \rho^2 & 0 \\ -\frac{2Mra\sin^2\theta}{\rho^2} & 0 & 0 & \frac{(r^2+a^2)^2 - a^2\Delta\sin^2\theta}{\rho^2}\sin^2\theta \end{pmatrix}$$

where

$$a = \frac{J}{cM}$$

$$\Delta = r^2 - \frac{2GMr}{c^2} + a^2$$

$$\rho^2 = r^2 + a^2\cos^2\theta$$

This contains a rather surprising feature: there are non-zero components outside the diagonal of the metric! These terms are directly proportional to a, which is directly proportional to the angular momentum J of the black hole. (Notice also that, when $a = 0$, this metric reduces to the Schwarzschild metric.) These non-diagonal terms represent frame dragging: if an object follows a geodesic path within a particular reference frame, then that reference frame is forced (through the $g_{t\phi}$ terms) to rotate around the source of mass in the same direction that the mass is rotating.

Imagine a particle which is initially at rest, such that $p_\phi = 0$. This has no angular momentum in its one-form. If we look at the four-vector describing its motion, we find that

$$p^\phi = g^{\phi\mu}p_\mu = g^{\phi\phi}p_\phi + g^{\phi t}p_t = -\frac{2Mra\sin^2\theta}{\rho^2}p_t$$

As $p_t = -mc^2$ in the rest frame of the particle, we have

$$p^\phi = \frac{2Mmra\sin^2\theta}{\rho^2} \neq 0$$

indicating that a particle must move in this spacetime. The angular velocity ω is given by

$$\omega = \frac{d\phi}{dt} = \frac{g^{\phi t}}{g^{tt}}$$

Now, in the matrix forms, $g^{\mu\nu} = (g_{\mu\nu})^{-1}$, and it is the version with the lowered indices which we have defined. Thus,

$$g^{\phi t} = -\frac{g_{\phi t}}{g_{\phi\phi}g_{tt} - (g_{\phi t})^2}$$

and

$$g^{tt} = \frac{g_{\phi\phi}}{g_{\phi\phi}g_{tt} - (g_{\phi t})^2}$$

such that

$$\omega = -\frac{g_{\phi t}}{g_{\phi\phi}} = \frac{2Mra\sin^2\theta}{\left((r^2+a^2)^2 - a^2\Delta\sin^2\theta\right)\sin^2\theta}$$

Notice that $\omega = 0$ when $a = 0$, as with the Schwarzschild radius. For large enough J, a begins to dominate. As one approaches the event horizon of a rapidly rotating black hole, there becomes a region in which p^ϕ can never take on a negative value: even a photon which is incident in the opposite direction to the rotation of the black hole is forced to orbit around the black hole in the direction of the black hole's motion.

This phenomenon occurs around anything which has mass and angular momentum, including our Sun. As the Sun is a fluid, there is no simple expression for its angular momentum. It has been shown, however, that this frame dragging is exactly what is needed to account for the seeming incongruities in Mercury's orbit which were known before Einstein developed the theory of relativity. This proved to be one of the greatest successes of the general theory of relativity.

8.5   Hawking Radiation: The Process

It is known that conservation of energy can be violated for very short periods of time. In other words, the Universe can cheat if it doesn't get caught. Stephen Hawking was the first to publish an analysis of what happens when this occurs near an event horizon. The specific constraint on energy and time fluctuations comes from the Heisenberg uncertainty principle:

$$\Delta E \Delta t < \hbar$$

where $\hbar$ is a (small) known constant. We now know that energy and mass are equivalent. Thus, two particles with mass m which appear in a quantum fluctuation represent a certain minimum amount of energy. Normally, these particles collide with each other and annihilate before the amount of time

$$t = \frac{\hbar}{2mc^2}$$

has elapsed. What happens if one falls through the event horizon? The other cannot annihilate with it any more; even if it did fall through the event horizon, the r coordinate is now timelike, so they could never intersect in space until they reach the singularity. This may take more time than is indicated. Hawking showed (through the mathematical mechanisms of quantum field theory, which fall far, far beyond the scope of this text) that the particle which falls into the black hole represents a negative amount of energy. The other particle drains enough energy from the black hole to become a real particle, possibly escaping the gravity well completely. The net result for the black hole is a loss of energy. The smaller the black hole, the faster this process will take place. Black holes with mass around that of stars will survive longer than the Universe, but black holes with small masses evaporate very rapidly. This is the process that terrifies the partially informed about particle accelerators. Yes, if a black hole on those scales is created, it will evaporate quickly with a high energy intensity. The total energy, however, is still tiny; it only has the energy it was created with. As particle accelerators ensure such collisions occur within particle detectors, there is nothing to worry about; this burst of energy will take place at a location in space and time that is within a device designed specifically to safely contain energy on those scales or greater. This doesn't even touch on the fact that the conditions to create those black holes are so unlikely that they probably will not happen within the lifetime of the hardware anyway.

8.6   Einstein-Rosen Bridges

In 1935, Albert Einstein and Nathan Rosen studied the possibility of Schwarzschild geometry existing without the crush of a black hole. They developed the concept of an Einstein-Rosen bridge, commonly known as a wormhole. The concept is simple: if a black hole punches a hole in space time, and spacetime can curve around on itself, then what happens if two black holes line up on two sides of space time? One could facilitate warp drive by going a shorter route through two black holes, right?

Wrong.

The two black holes in question can only connect in this manner if certain conditions are met:

- They must be connected at the moment of creation.

- Their geometry must be identical.

- They must be perfectly aligned right from the start. If they rotate, the angular momentum must have identical magnitudes but opposite directions to maintain this alignment.

These conditions are so exacting that we cannot expect wormholes to be a naturally occurring phenomenon. Identical black holes with identical mass are hard enough to construct, but when you also factor in the simultaneous creation and perfectly planned positioning, it is clear that this will not happen at random. This, in itself, does not eliminate the possibility of warp drive; none of these conditions prohibit perfectly planned artificial construction. That comes into play from the final condition: the geometry must remain identical. If you drop in a mass m in one side of the wormhole, you must drop an identical mass m in the other side at a precisely identical moment. These must be identical in all respects, so if you are dropping in a space vehicle with mass m (including crew) you must have a mass with identical geometry on the other side, which effectively means a ship made out of clones of members of the original ship who are so identical that they would have the same brain patterns and would make all the same decisions with all the same resources.

To this point, it is all possible, though extremely unlikely. The difficulty come in examining the trip: despite some popular science fiction, there is no such thing as a "white hole" on the other side spewing things out. What falls into the black hole stays in the black hole. If you align everything perfectly so that you enter the wormhole without breaking it, you still end up getting dragged inexorably towards the singularity, completely unable to escape. In other words, your experience falling into a wormhole will be *identical* in every respect to the experience you would have falling into a black hole. This is not warp drive: this is astonishingly complicated and expensive suicide.

Chapter 9

Cosmology

9.1 What Is Cosmology?

Cosmology is the study of the structure and nature of the universe. Instead of focusing on any single object, cosmologists study the Universe as a whole and the way it evolves over time.

9.2 Einstein's Cosmological Constant

With the discovery of the general theory of relativity, the geometry of the universe was quantifiable and definable. This meant that, for the first time, the structure of the Universe could be studied explicitly. Einstein was amongst the first to do this calculation, and he ran into a result that disturbed him greatly. When he examined the geometry of the Universe we live in, he uncovered math that made it appear as though the Universe was expanding. This did not fit his ideas of theology: Einstein felt that the God he believed in would have created a perfect Universe to begin with, and expansion represented an undesirable form of change. The equations included a constant whose value could not have been measured at the time. Einstein arbitrarily chose a negative value for this constant which would result in a static Universe which would prevent this predicted expansion.

A few short years later, Edwin Hubble published his results which proved that the Universe was, in fact, expanding. This shocked Einstein and others. The Universe was an evolving beast of some form. Einstein and others recognized his earlier error and made another one: they arbitrarily decided that the cosmological constant should have a value of zero. This led to an expanding Universe model, and everything seemed fine. It seemed clear that, at some point in the past, the Universe had been a finite point, which then exploded in a Big Bang, and began expanding from that point.

This was the accepted model for almost 80 years. It was challenged in 1998, when a group of scientists recognized that the value of zero is just as arbitrary and any other value for the cosmological constant, so they set about trying to measure the value explicitly. What they discovered shocked just about everybody: the cosmological constant is not zero. Not only that, it is a positive number. In other words, the Universe would expand whether it contained matter or not. The question is: why? What physical process or object does this cosmological constant represent?

This is still one of the unanswered questions of cosmology. There have been a number of proposals, most of which involve dark matter or dark energy. Astronomical observations show expansion behaviour that cannot be accounted for by the matter and energy we can observe. Therefore, it has been proposed that there is some sort of matter or energy out there which we cannot observe, which is commonly referred to as dark matter and/or dark energy. Very little is known about dark matter, although much has been suggested. The difficulty with studying dark matter is that it is dark; it cannot be directly observed or studied using today's technology. This may change in the future as new technologies are developed, but that day still appears to be a long way off.

9.2.1   Expanding Into What?

One of the philosophical questions that has been asked is what the Universe is expanding into. Is there a larger universe with greater dimensions outside of this one? If so, what is that Universe "in?" Is there an infinite chain of universes, like some infinite set of matryoshka dolls?

The question about what is outside our Universe, if anything, will probably never be answered. Our instruments are limited to measuring that which is in our Universe, and that alone. It has been shown, however, that it is possible for the complete "exterior" geometry of such a Universe to remain unchanged, while the "internal" geometry continues to expand. If there is an outside Universe, our presence in that Universe may not appear to be changing at all.

9.3   The End(s) of Time

One of the most commonly asked questions in cosmology is "how will the Universe end?" There have been a number of proposals for this.

- **The Big Crunch** - In this theory, the expansion of the Universe eventually slows, and the Universe starts to fall back in upon itself, destroying all life in the process.[1] This theory has a companion theory, in which the lifetime of our Universe is cyclic, and the Big Crunch would be followed by a new Big Bang, with a new Universe being formed. This would happen over and over again in an infinite cycle.

- **The Static Universe** - This theory proposes that there is just enough mass around to slow the expansion of the Universe without ever quite halting it. This, ultimately, leads to a Universe very much like the static Universe Einstein's theological instincts wanted. It wouldn't technically be static, but the

---

[1]Unless you are a being named Galan who likes to snack on planets. Then you might survive.

expansion would eventually become imperceptibly slow. In this case, thermo-dynamics predict that all energy would eventually be converted into unrecov-erable waste heat due to entropy, wiping out all life in the "heat death" that results from all of existence boiling away into a plasma that no longer allows complex molecular structures to exist.

- **The Constantly Expanding Universe** - In this theory, the Universe simply continues to expand, causing an increasing space between civilizations and worlds, and reducing the average energy in any given volume. Life, in this model, would eventually die off when all temperatures in the Universe drop below the point at which chemical reactions can occur and all of existence is frozen out at temperatures very near absolute zero.

The discovery of a positive cosmological constant make the third theory the most likely theory. Tell your descendants to bundle up: it's going to get cold.

9.4 The Shape of the Universe

The simplest cosmology to work with in general relativity is a combination of the work of Einstein, Robertson and Walker. Robertson and Walker added two assump-tions to Einstein's field equations:

1. At any instant in time, the entirety of space appears homogeneous and isotropic, barring the effects of mass an energy within it. In other words, any two loca-tions in an empty universe are utterly indistinguishable from each other.

2. Our definition of simultaneity makes sense from the perspective of this uni-verse.

These assumptions ultimately result in a metric that gives the line element

$$dl^2 = \frac{dr^2}{1 - kr^2} + r^2 d\Omega^2$$

where $d\Omega$ is the unitless volume element of the spatial components of the spacetime, and k is defined by

$$g_{rr} = \frac{1}{1 - kr^2}$$

as a representation of the spatial dependence of the curvature of spacetime, relative to the "centre" of this cosmology.

The assumptions provided do not allow us to specify a value for k alone. It is here that the different possibilities for the future of the Universe appear; the value of k determines the shape of the cosmology.

In the case in which $k = 1$, the spatial extent of the universe at a given point in time is a perfect sphere. This is a closed spacetime; the universe folds back in on itself, indicating that a sufficiently fast and long-lived individual can travel along a geodesic, circumnavigate the universe, and return to his or her starting point, just as a person can walk (and/or swim) in the straightest line possible starting at Earth's North Pole and return to that North Pole later. Other positive values of k also result in closed universes, but they are not perfectly spherical. At the point $k = 0$, the universe is flat Minkowski space as in the world of special relativity. Finally, if $k = -1$, the universe is a perfect hyperbolic shape, never ending or closing in upon itself in what is called an "open" geometry.

There are certainly other cosmologies, but most of them involve assumptions which are less intuitive. That doesn't mean they are incorrect, but it does mean further evidence will be required before one can find a good candidate amongst them to describe our universe.

9.5   The Fate of the Universe

The k parameter described above is one of the two parameters that are significant when it comes to describing the ultimate fate of the Universe. The second is Einstein's famous, or possibly infamous, cosmological constant.

To see how the cosmological constant arises, we need the stress-energy tensor. We can treat spacetime as a fluid, which is pulled, pushed and twisted around by gravity. This can be described by the stress energy tensor $T^{\alpha\beta}$ which describes the flux of momentum p^{α} over a surface of constant x^{β}.[2] This definition and the Ricci tensors are enough to show that the ultimate gravitational fields $G^{\alpha\beta}$ are related to the metric and stress energy tensor by

$$G^{\alpha\beta} = kT^{\alpha\beta} - \Lambda g^{\alpha\beta}$$

where Λ is a constant of integration that appears along the way. Einstein set this constant, now known as Einstein's cosmological constant, to a specific value that created a "steady state" universe, meaning that the Universe would not expand. After Hubble's discovery, the scientific community started treating the equations as though $\Lambda = 0$, but that choice is just as arbitrary. In fact, recent attempts to explicitly study and measure this constant indicate that it is not only non-zero, but that it has a positive value sufficient to force the expansion of an empty universe.

Nonetheless, it is the combined values of k and Λ that ultimate determine the fate of the universe as it expands or contracts for the rest of time. At this time, we do not have sufficient empirical evidence to determine which possibility is the real one, or even of the Robertson-Walker cosmology is the correct one; some of those exotic assumptions leading to other cosmologies may turn out to be correct.

---

[2]See any fluid dynamics course for complete details of how to define this.

Bibliography

[1] Albert Einstein. *Relativity*. Routledge, 2013.

[2] Isaac Newton. *The Principia: mathematical principles of natural philosophy*. Univ of California Press, 1999.

[3] Bernard Schutz. *A first course in general relativity*. Cambridge university press, 2022.